ELECTRIC UTILITY MERGERS

ELECTRIC UTILITY MERGERS

Principles of Antitrust Analysis

Mark W. Frankena
and
Bruce M. Owen

with contributions by
John R. Morris and Robert D. Stoner

PRAEGER

Westport, Connecticut
London

Library of Congress Cataloging-in-Publication Data

Frankena, Mark W.
Electric utility mergers : principles of antitrust analysis / Mark W. Frankena and Bruce M. Owen with contributions by John R. Morris and Robert D. Stoner.
p. cm.
Includes bibliographical references and index.
ISBN 0–275–94596–0
1. Electric utilities—United States—Mergers—Case studies.
2. Antitrust law—United States. I. Owen, Bruce M. II. Title.
HD9685.U5F63 1994
338.8'3613337932'0973—dc20 93–36702

British Library Cataloguing in Publication Data is available.

Library of Congress Catalog Card Number: 93–36702
ISBN: 0–275–94596–0

First published in 1994

Praeger Publishers, 88 Post Road West, Westport, CT 06881
An imprint of Greenwood Publishing Group, Inc.

Printed in the United States of America

The paper used in this book complies with the Permanent Paper Standard issued by the National Information Standards Organization (Z39.48–1984).

10 9 8 7 6 5 4 3 2 1

Contents

Figures and Table

FIGURES

TABLE

Preface

Competition in the generation, transmission, and distribution of electricity is of increasing interest to policy makers. The use of competition as a social policy tool to benefit consumers carries with it the necessity to preserve competition that is threatened by mergers or other structural changes. In this book, we set out to explain the central principles of antitrust economics and their application to the analysis of proposed electric utility mergers.

Our target audience includes utility executives and users of electric power and transmission services who may be considering the economic effects of proposed mergers, as well as the lawyers and economists who work on their behalf. Our audience also includes members of utility commissions and their staffs, members of the judiciary, and other government officials who must decide whether proposed transactions are likely to bring benefits or harm to the public. We also hope that our book will be useful to students in law, economics, and business courses dealing with antitrust, industrial organization, and utility regulation.

We want to be frank about our motivations for writing this book. We came to study electric utilities in 1989, when we were retained by the city of San Diego in connection with a proposed merger between two giant California utilities. The city was interested in opposing the merger. Lawyers for the city believed that if antitrust experts examined the facts, they would conclude that the merger was anticompetitive. As it turned out, we did conclude that, and hearing officers at both the federal and state levels, the California Public Utilities Commission, and lawyers at the U.S. Department of Justice agreed with that assessment. The proposed merger was abandoned after it was rejected by the California commission. Subsequently, we testified on behalf of Public Service of New Hampshire on its merger with Northeast Utilities and on behalf of Occidental Chemical Corporation on the proposed merger of Entergy and Gulf States Utilities.

In conducting our investigations of these mergers, and in participating in the hearings, we were struck by the contrast between the level of sophistication regarding competitive effects with which mergers are analyzed at the Department of Justice and the Federal Trade Commission (organizations with which we had been associated) and the analyses of the staffs and the parties at the regulatory commissions dealing with electric power. Our first motivation for writing this book was to encourage the appropriate analysis by regulators and company representatives of antitrust issues that arise in electric utility mergers.

While this book focuses on mergers, the economic analysis explained here also will be useful in analyzing many other issues that are becoming increasingly important with the growth of competition in electric power. For example, proper definition of markets and analysis of market power will be useful in decisions on whether to continue regulating certain transactions.

Our second motivation for writing this book is a more delicate matter. In reading the literature on the electric power industry as part of our investigation of the California utility merger, we were struck by the fact that much of that literature is the product of experts who are regularly retained by private utilities in connection with regulatory proceedings or litigation. These experts have had access to facts and analyses that, over the years, have reinforced their expertise. Opportunities to study an industry and to become an expert are greatly enhanced by access to the decision makers in the industry and to the private data of its firms. The environment of litigation is conducive to the collection of very large bodies of facts for studies that probably could not be afforded for purely scholarly ends or with purely academic means. Consulting work and scholarly research, therefore, can be mutually reinforcing. The expert is sought out by advocates on behalf of parties in litigation in part because decision makers are likely to give weight to the credentials of the expert.

In spite of the beneficial reinforcing effects of scholarly research and consulting work, there are dangers. Among them is the fact that the leading experts tend to be hired by the parties with the most money. This does not mean that the experts' opinions are for sale! Experts, academic or otherwise, often turn down invitations to participate on one side of a case or another because they disagree, or expect after investigation to disagree, with the position advocated by that party. When an expert turns down an offer to work for one side, however, that does not mean that the other side will have any interest in (or resources to spend on) hiring that expert. The "other side" from an electric utility is often an impecunious state regulatory commission staff or an equally poverty-stricken municipal distribution system. The best experts are in short supply. Competition drives up their prices. They have more than enough opportunities to occupy their time working for parties with whose positions they agree. Because consulting experience reinforces scholarly research, we think consulting experience largely limited to working for utilities may produce a somewhat different research product than would a different set of experiences.

We do not want to suggest that there is anything sinister or underhanded about this process. On the contrary, we think the process is on the whole healthy and beneficial in that it expands the amount, depth, and relevance of research. In many industries there are more than enough well-funded "sides" and experts to have varied perspectives. In the electric utility business, however, there is an imbalance in the consulting experience of the leading academic experts, as there was in the telephone industry before the breakup of AT&T.

In writing this book we have been alert to the need to provide a fresh perspective. We bring to this task a relative lack of firsthand experience in the electric power industry that is, for the reasons just explained, both a handicap and an advantage. Our backgrounds are primarily in antitrust economics and industrial organization. Bruce Owen was chief economist of the Antitrust Division of the U.S. Department of Justice. Mark Frankena was Deputy Director for Antitrust in the Federal Trade Commission's Bureau of Economics. Robert Stoner, who wrote Chapter 9, was Deputy Assistant Director for Antitrust at the Federal Trade Commission. John Morris, who co-authored Chapters 1, 4, and 5, was Assistant to the Director for Antitrust at the Federal Trade Commission. Owen is now president of Economists Incorporated, a Washington, D.C., consulting firm where Frankena, Morris, and Stoner are senior economists.

We are grateful to Heather Grace, Justin Lu, and Tom Wilson for research assistance, to Muriel Warren, Cindy Walton, and Renee Witcher for secretarial assistance, and to Barbara de Boinville, Fran Kianka, and Sally M. Scott for editing the manuscript.

ELECTRIC UTILITY MERGERS

1

Market Power and Antitrust Analysis

Antitrust policy is motivated by the substantial benefits of competition. A preference for competitive markets is built into the economic policy of the United States. In 1890 the Sherman Antitrust Act codified a common law tradition that frowned on monopolies and restraints of trade. No other developed country has had as much interest in or experience with antitrust policy as the United States. Recently, as the rest of the world has moved toward private ownership and less regulated market-oriented economies, many of the principles of competition law that are practiced in the United States have been adopted abroad.

There are many reasons for choosing competitive market solutions to the economic problem of determining how much of which goods to produce for whom. First, competition tends to allocate resources efficiently. Scarce and therefore valuable resources are assigned to their most productive uses, and the resulting output is allocated to the customers who value it most.

Second, competition provides powerful incentives for producers to minimize the cost of production, that is, to avoid waste. Producers seek profits. Finding ways to cut costs is one way of making a profit. Firms that are lazy and backward in their production techniques will lose money and be driven from the market as more efficient producers cut prices to make sales. Only the efficient producers will survive.

A third advantage of competition is that producers are rewarded when they are responsive to consumer needs and are punished when they are not. Sellers that are unresponsive are abandoned by their customers.

Competitive markets serve consumers' interests. This is the major justification for policies that promote competition. In addition, competition is consistent with widely held political ideals. Competitive markets diffuse economic power. Important decisions are taken by market forces rather than by the arbitrary voices of powerful economic interests or central planners and bureaucrats. Adam Smith's "invisible hand" holds sway rather than some economic or political oligarchy.

This book is about the application of antitrust analysis to electric utility mergers. No one can hope to apply this analysis without an appreciation for the reasoning that lies behind our legal preference for competitive market processes.

THE COMPETITIVE PROCESS

A "perfectly competitive" market would have many buyers and many sellers, each taking prices as given. In such a market, prices adjust so that what consumers are willing to buy usually matches what producers are willing to sell. Firms produce what consumers want, neither withholding output to increase prices nor wastefully overproducing. Unexpected increases in demand lead to higher prices, giving some producers attractive profits. These producers, and perhaps new producers, expand output to capture the higher profits. The additional output lowers prices until what consumers are willing to buy once again matches what producers are willing to sell. Unexpected increases in supply lead to lower prices, so some producers lose money. These producers decrease their output or exit from the business to reduce their losses. The reduced output causes prices to rise to a level where consumers once again are willing to buy what producers are willing to sell. Prices of individual goods and services rise or fall in reaction to changes in supply and demand.

Prices and profits or losses serve as important "signals" in any competitive market. High prices result in temporarily high profits, calling forth increased production and entry; this directly benefits consumers by giving them more goods at lower prices. Low prices lead to losses, production cutbacks, and exits; this indirectly benefits consumers by eliminating the wasteful use of resources in the production of unwanted goods.

In more complicated circumstances the market process works in a similar way. For example, in many markets, decisions concerning plant location or product quality are as important as decisions about output levels. As long as there are many buyers and many sellers, each too small individually to affect market outcomes, the "invisible hand" will tend to translate the private greed of sellers into the public good of efficient resource allocation, maximizing the overall wealth of society.

Consumers are the primary beneficiaries of competitive markets. When markets are competitive, consumers' economic well-being is maximized, somewhat ironically, by producers' pursuit of profit. Effects on consumers also provide a central test for whether an action—such as a merger transaction—is anticompetitive. Many activities are ambiguous and difficult to interpret under the antitrust laws. Clarity can often be achieved by considering whether the activity tends to promote the economic interest of consumers as a group.

MARKET IMPERFECTIONS

A number of things can go wrong with a market, with the result that the pursuit of private profit may not serve the interests of consumers. Most of these market imperfections are of no immediate concern in the antitrust analysis of electric utility mergers. Consider, for example, the problem of externalities. Competitive markets may break down if the activities of producers or consumers affect others through nonmarket channels. When factories emit smoke, residents breathe polluted air. The factory takes no account of the costs imposed by this nuisance. If it did, it would emit less smoke and produce less output. When a resident cuts her lawn, the noise of the mower disturbs others in the neighborhood. The lawn-mowing neighbor may take no account of this cost of her activity. If she did, she would mow less or at a different time. These external effects of production or consumption activities reduce the efficiency of market processes unless they are brought within the market through the more complete assignment of property rights, including rights to air quality and peace and quiet.

Antitrust law is concerned with a different market imperfection: the problem of market power. A *monopolist* is a single seller of a good or service for which buyers have no good substitutes at competitive prices. Other firms are unable to supply this good or service profitably. Such a seller is said to have *market power*. This means that the monopolist can make extra profits by raising prices above competitive levels, up to the point where substitute goods begin to look attractive to consumers, or where potential rivals could profitably enter the market. A *cartel* is a group of sellers acting cooperatively, or colluding, so that their pricing behavior resembles that of a single seller. Cartels or price-fixing agreements present the same economic problems as monopolies. Cartels, however, may behave in ways that a simple monopolist would not, because a cartel needs to detect and deter cheating on the cartel agreement by members, which individually have an incentive to increase their sales. Mergers are of concern in the context of antitrust analysis

Figure 1.1: Monopoly Pricing

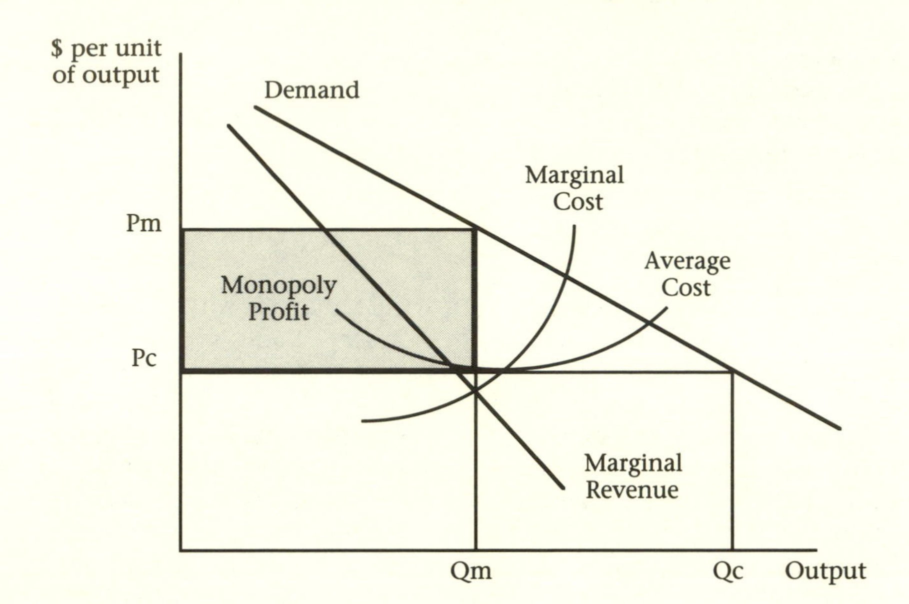

because they are one means by which market power may be created or enhanced.

Figure 1.1 illustrates the problem of monopoly pricing. The height of the Demand line at any output level expresses what consumers are willing to pay for an additional unit of this service. The height of the Marginal Revenue line shows the increase in the monopolist's total revenue that would result from selling an additional unit of output, allowing for the fact that to sell additional output the monopolist must reduce the price at which all units of its output are sold. The Average Cost curve represents the production cost per unit of output, and the Marginal Cost curve represents the incremental cost of producing an additional unit of output. These costs include a competitive rate of profit.

If the industry were competitive, prices would settle at about P_c and output would be Q_c. The price, P_c, would be equal to both Average Cost and Marginal Cost, and thus firms would make only a competitive rate of profit. In contrast, the monopolist charges a price of P_m in order to maximize profits, and produces an output of Q_m, which is where Marginal Revenue equals Marginal Cost. The monopolist produces less than the competitive output because the monopolist takes into account the fact that to sell additional output prices must be reduced. The additional revenue earned by the monopolist from increasing sales from Q_m to Q_c, which would reduce the price from P_m to P_c, would be less than the incremental cost of producing the extra output.

The monopolist's decision to charge P_m has several consequences. First, compared with the competitive situation, too little of this service is produced and consumed. Second, there is a transfer of wealth from consumers to the monopolist (equal in quantity to the "Monopoly Profit" rectangle in the diagram). Third, there is a "deadweight loss" of economic wealth (equal in quantity to the triangular area to the right of the Monopoly Profit rectangle and to the left of the Demand line). This represents a loss to consumers for which there is no corresponding benefit to the monopolist.

Figure 1.1 is drawn to illustrate an "unnatural" monopoly: one in which cost conditions do not mandate a single seller. In the case of electric utilities, it is frequently assumed that there are elements of "natural" monopoly. This means that the height of the average cost curve declines as output increases over a large range of output, so that

Figure 1.2: Natural Monopoly

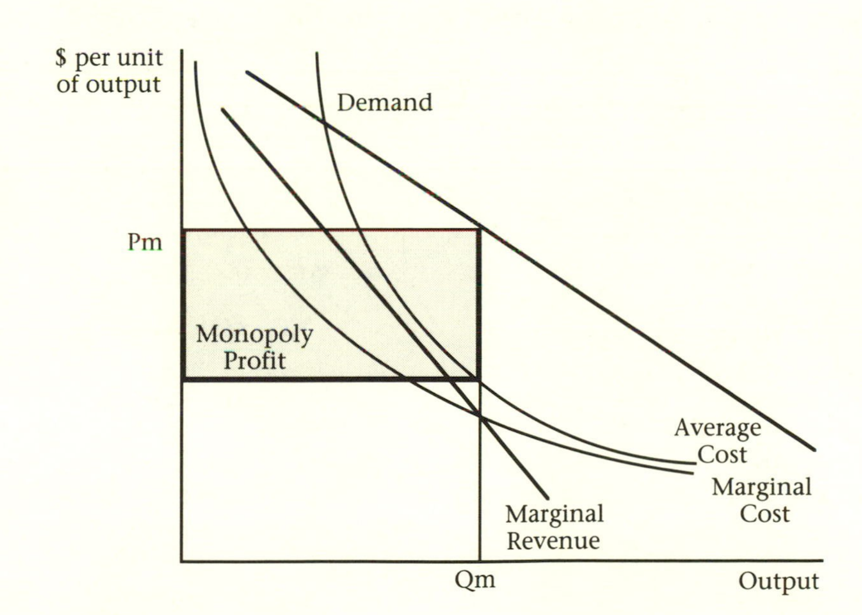

one producer can serve the whole market at lower cost than can two or more firms (see Figure 1.2).

Whether electric utilities are natural monopolies is subject to dispute. Even if they are natural monopolies, the principles of competition and antitrust law remain applicable to many aspects of the operation and regulation of power companies, including mergers. The principles of competition analysis apply to a regulated monopolist's incentives in making output and pricing decisions and in dealing with both customers and suppliers.

Monopoly can exact an additional cost penalty. A monopolist that is protected from the threat of entry has less need to be efficient in its production processes and may take some of its profits in the form of a quieter, less entrepreneurial life. In Figures 1.1 and 1.2, the cost curves were drawn to represent the *minimum* cost at which output could be produced. In the real world, monopolists may seldom have the incentive to cut this close to the bone. Their effective cost curves may lie above the minimum cost of production.

The inefficient monopolist illustrated in Figure 1.3 imposes a cost on society of productive inefficiency. It does not employ the least costly method of production. The curves labeled "MC - Monopolist" and "AC - Monopolist" are based on the assumption that the monopolist's mar-

Figure 1.3: Monopoly That Does Not Minimize Costs

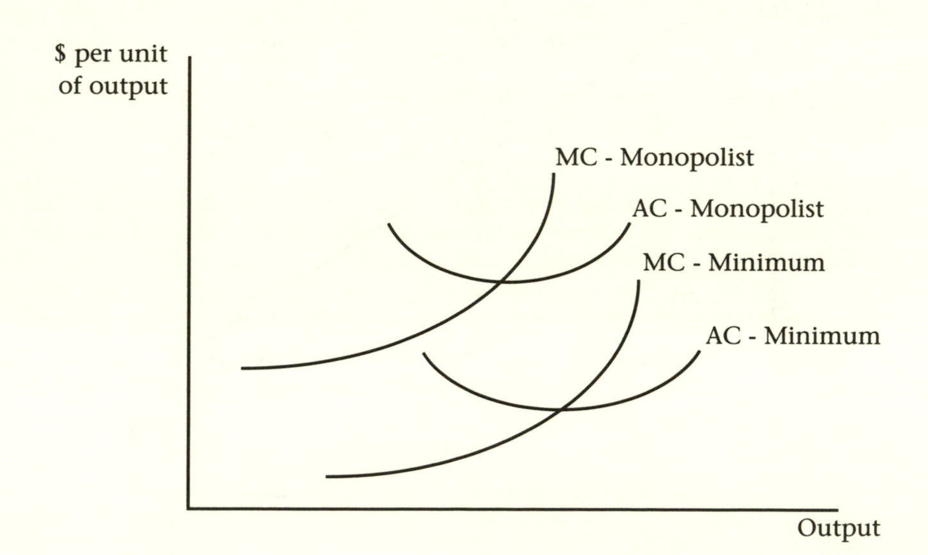

ginal and average costs are above the levels that would prevail with competition, which are represented by "MC - Minimum" and "AC - Minimum." For instance, when AT&T was broken up in 1984, the costs and prices for newly competitive telephone services and equipment plummeted. In contrast, the local telephone companies that remained monopolies raised their prices.

Finally, a monopolist earns profits in excess of those required to cover its costs, including capital costs. These "monopoly rents" are worth fighting for. In fact, it would make sense for a firm with monopoly rents to pay up to the amount of those rents in expenditures to preserve its monopoly. Such expenditures are sometimes called "rent-seeking" activities. They include the efforts of lobbyists, lawyers, and consultants hired to facilitate the retention of government privileges such as monopoly franchises, the imposition of government restraints on rivals and potential rivals, and the evasion or avoidance of regulatory constraints on profits. Because these expenditures are undertaken largely to influence government action, the existence of a monopoly creates a political force that would not exist in a competitive market. (Competitive markets are not, of course, politically neuter. Often competitive producers such as farmers are able to form rent-seeking coalitions to obtain, for example, tariff protection.)

Any monopoly rents earned by an electric utility at the expense of consumers are not necessarily its to keep. Labor unions, managers, and interest groups such as environmentalists all lay claim to portions of these rents. In the end, stockholders may enjoy few of the fruits of their monopoly.

ANTITRUST LAW

Antitrust law in the United States, although founded on statutes, is mainly made by judges. The statutes themselves are rather vague and their goals are facially ambiguous. One must read the judges' decisions to understand the meaning of the law.

There are two broad categories of antitrust law: behavioral law and structural law. *Behavioral law* seeks to prevent competing firms from agreeing on prices or other terms of trade. In addition, this branch of the law seeks to control certain unilateral practices that might extend or enhance a firm's market power. *Structural law* seeks to prevent the accumulation of undue market power through industrial consolidations, such as mergers. In addition, this branch of the law deals with extreme cases of monopolization, where remedies such as dissolution or divestiture may be appropriate.

In the United States the three principal statutory bases for antitrust law are Sections 1 and 2 of the Sherman Act and Section 7 of the Clayton Act. Section 1 of the Sherman Act is the basic anticollusion provision of the antitrust laws:

> Every contract, combination . . . , or conspiracy, in restraint of trade or commerce . . . is declared to be illegal. Every person who shall make any contract or engage in any combination or conspiracy hereby declared to be illegal shall be deemed guilty of a felony, and, on conviction thereof, shall be punished by a fine not exceeding $10,000,000 if a corporation, or, if a person, $350,000, or by imprisonment not exceeding three years, or by both. (26 Stat. 209 (1890) (amended 1990), 15 U.S.C.A. §1 (1993))

As interpreted by the courts, Section 1 prohibits price-fixing and other cartel-like actions. In addition, some supplier-customer relationships (such as agreements on the resale prices of goods) are prohibited.

Section 2 of the Sherman Act prohibits anticompetitive attempts to monopolize and willful maintenance of a monopoly position: "Every person who shall monopolize, or attempt to monopolize, or combine or conspire with any other person or persons, to monopolize any part of the trade or commerce among the several States, or with foreign nations, shall be deemed guilty of a felony" (26 Stat. 209 (1890) (amended 1990), 15 U.S.C.A. §2 (1993)). Section 2 does not make all monopolies unlawful. The act gives the courts the power to impose fines and prison terms and to order divestitures and dissolutions.

Section 7 of the Clayton Act, the principal merger law, condemns mergers or acquisitions the effect of which "may be substantially to lessen competition, or to tend to create a monopoly" (38 Stat. 730 (1914), 15 U.S.C.A. §18 (1993)).

Following are other important features of the federal antitrust laws:

- Establishment of two separate agencies, with largely overlapping jurisdictions, to enforce the antitrust laws. The Antitrust Division of the U.S. Department of Justice is a prosecutor that brings its criminal or civil complaints before federal judges. The Federal Trade Commission is an administrative agency with its own internal judicial system. It does not deal with criminal cases.
- Provisions for a system under which large mergers or acquisitions must be reported to the government in advance, so that the Antitrust Division or the Federal Trade Commission can seek to stop anticompetitive transactions before they are consummated (Clayton Act §7A).
- Prohibitions on certain pricing practices, such as tying and price discrimination.
- Provisions for private persons injured by those who violate the antitrust laws to be awarded treble damages (Clayton Act §4).

The behavioral and the structural branches of antitrust law are linked in important ways. The principal reason for concern about concerted activity among competitors is that collective behavior may resemble the behavior of a monopolist. The reason for concern about certain unilateral activities of a monopolist or other firm with market power is that the firm's market power will be strengthened or its exercise of market power perfected. One reason for seeking to control mergers and acquisitions is that competitors should not be allowed, through merger, to achieve anticompetitive ends they cannot lawfully achieve through agreements. In addition, mergers may increase the likelihood of successful collusion among remaining firms.

In each of these cases, the goal is to prevent an increase in price, a reduction in output, or the equivalent in nonprice dimensions, such as a lowering of quality. The underlying assumption in each case is that violation of the law results both in a deadweight loss of economic welfare and a transfer of wealth from customers to producers. The different branches of antitrust law thus reflect a common concern with the creation and exercise of market power.

A unique feature of U.S. antitrust law is the private treble damage action. When combined with the possibility of a class action, this is a powerful tool for customers injured by price-fixing agreements among suppliers. The deterrent effect of such action protects all consumers. At the same time, it is often suggested that the potential windfall for plaintiffs and their attorneys provides an incentive for private suits of dubious merit.

U.S. antitrust law has at times been used to restrict competition rather than to promote consumer welfare. This has been true of so-called vertical price-fixing cases, which are often complaints by distributors against manufacturers or franchisers. It has also been true in the case of complaints by competitors of merging firms. Today, however, it is generally recognized that the antitrust laws exist for the "protection of *competition*, not *competitors*."[1]

From the point of view of enforcement of the antitrust laws, it is useful to think of a spectrum of potentially anticompetitive activities ranging from ones that are almost certain to injure competition to those whose effects on competition and consumer well-being require careful analysis. Linked to this spectrum of potential offenses is a system of increasingly demanding judicial standards of liability. For example, conspiracies among competitors to fix prices or to restrict output are regarded as so clearly pernicious, and so seldom capable of any serious justification, that they are condemned out of hand as per se illegal. Such activities may not be defended on the grounds that the agreement or its results were reasonable. Most other potential offenses require analysis and can, at least in principle, be defended successfully on the ground that the alleged offense had substantial benefits for consumers. Such "rule of reason" cases that are based on a cost-benefit comparison comprise the bulk of the noncriminal caseload.

THE ROLE OF REGULATION

Public utility regulation can promote consumer benefits and efficient use of resources in some situations where antitrust policy cannot. Antitrust policy cannot correct most problems created by monopoly. It does not condemn monopoly itself or even the charging of a monopoly price. It condemns only certain means of acquiring, maintaining, or exercising market power. In some situations, either production costs, laws, or regulations may dictate that only one firm or a very limited number of firms can supply a product. The distribution of electric power to residential users is an example. With only one potential supplier, the necessary conditions for competition are not met, and traditional antitrust intervention will not prevent the seller from charging monopoly prices. In such cases, direct regulation of prices and services may benefit consumers by lowering prices toward marginal costs and ensuring service closer to competitive levels.

Where competition is possible, however, regulation is at best an imperfect substitute. Regulation is costly to taxpayers and to regulated firms. Yet regulation cannot be expected to result in the prices that would result from competition, since regulators have limited information and other resources. Inevitably, regulators will permit the exercise of market power in many cases, and they may deter efficient investments in other cases.

Some of the imperfections of regulation arise because of deficiencies in the rules applied by regulators. An example is the allocation of joint production costs among services or customers. Suppose that a utility has one power plant that serves residential and industrial customers. Various prices would be consistent with the allowed rate of return. Regulators may decide to allocate the fixed cost of the plant between the two customer groups based on unit sales. The resulting prices, however, are not likely to be the ones that would lead to the most efficient use of electricity by residential and industrial customers. The efficient set of prices would allocate costs based on each group's responsiveness to the price of power, not its consumption level. The group with fewer substitutes for electricity, and therefore lower responsiveness to the price of electricity, can be charged a higher price without causing as much distortion in its consumption of electricity and other products.

Thus it might be efficient for residential customers to pay for a majority of the fixed costs of the power plant even if they purchase a minority of the power.

To the extent that regulators are successful at lowering prices and expanding output toward competitive levels, they create strong incentives for utilities to evade the regulation. A regulated utility may evade cost-based regulation of its retail prices by paying an unregulated affiliate artificially high prices for power. The high price may be passed along to the customers, while the unregulated affiliate captures the monopoly profits. Vertical integration may also result in foreclosure of competition. A local distribution company may purchase power at a high price from an affiliate that has high generating costs even though lower-cost nonaffiliated suppliers would have been willing to sell the power at lower rates.

Rate-of-return regulation may create an incentive to own too much capital when the allowed rate of return exceeds the utility's cost of capital. By increasing the rate base, the utility may increase the total amount of profits that it may keep (Averch and Johnson 1962). For example, phone companies may have required customers to rent telephones to expand the phone companies' rate bases. By having a large stock of telephones in the rate base, the companies were able to earn greater profits than if customers owned phones.

GROWING COMPETITION IN THE POWER BUSINESS

Changing market conditions and regulations are leading to greater competition in the electric power industry. Deregulation and increased competition in the supply of natural gas have created additional energy supplies to compete with electricity. Additional gas supplies and improvements in small-scale natural gas combined-cycle and combustion-turbine technology have made nonutility power production more attractive. Large power units with long lead times have become more risky because of growing uncertainty about demand.

Regulation of the electric power industry—like regulation in the transportation, telecommunications, and natural gas industries—has changed to promote greater reliance on competition, particularly in generation. Growth of nonutility generating companies has been promoted in part by developments that overloaded the traditional regulatory process. From the early 1970s to the early 1980s, rate regulation prevented utilities from raising their prices as fast as their costs. Beginning in 1985, regulators prevented substantial amounts of utility investment from being included in rate bases. In addition, nonutility generation was encouraged by the Public Utility Regulatory Policy Act of 1978, which enabled cogenerators and small power producers—known collectively as "qualifying facilities"—to avoid traditional state and federal regulation and to sell their power to local utilities on favorable terms. The Energy Policy Act of 1992 greatly reduced entry barriers for nonutility generating facilities designed to sell wholesale power. This act also gave the Federal Energy Regulatory Commission power to mandate provision of wheeling services (the transmission of bulk power owned by other entities), in an effort further to increase competition in generation. Greater competition has increased the scope for mergers to have both procompetitive and anticompetitive effects, and hence the incentives for electric utilities to merge.

ANTITRUST REVIEW OF MERGERS

Any substantial merger or acquisition in the electric power industry goes through an intensive review process. A transaction is likely to be reviewed by several federal agencies, a number of state agencies, and private parties, any number of which may consider the principal issues in an antitrust analysis: competitive effects and cost savings.

Although subject to the antitrust laws, electric utility mergers are mainly reviewed by the Federal Energy Regulatory Commission and state utility commissions. These investigations are quite different from those conducted for other mergers by the Antitrust Division of the U.S. Department of Justice and by the Federal Trade Commission. First, the statutes under which utility mergers are evaluated by utility regulators use public interest standards that include but go beyond the competitive effects and efficiencies that are considered by the federal antitrust agencies. A federal appeals court recently held that the Federal Energy Regulatory Commission (FERC) is not required to use standard antitrust analyses, stating that, although antitrust considerations must be taken into account under the public interest standard, "There is no evidence that Congress sought to have the Commission serve as an enforcer of antitrust policy in conjunction with the Department of Justice and Federal Trade Commission" and that "FERC is not bound to use antitrust principles when they may be inconsistent with the Commission's regulatory goals (*Northeast Utilities Service Co. v. Federal Energy Regulatory Comm'n.*, No. 92-1165 (1st Cir. 1993)). Second, the analyses of market power carried out by the Federal Energy Regulatory Commission frequently display a lack of understanding of traditional antitrust analysis. Third, the procedures followed by utility regulators allow extensive participation by interested parties.

Federal Review

The Federal Energy Regulatory Commission reviews mergers under the Federal Power Act to determine whether they are consistent with the public interest. Its review includes competitive effects and cost savings. Either the Federal Trade Commission or the Antitrust Division of the Department of Justice could conduct an independent review to determine whether a merger is consistent with the antitrust laws. Recently, it has been the Antitrust Division rather than the Federal Trade Commission that has reviewed electric utility mergers. Rather than carrying out an independent investigation, the Antitrust Division has limited its role to (at most) participation as an interested party in the Federal Energy Regulatory Commission's review.

The Securities and Exchange Commission also reviews utility mergers for compliance with various securities laws and regulations, including the Public Utility Holding Company Act. In the case of mergers between what are defined under the act as "registered public utility holding companies," the Securities and Exchange Commission is required to consider price effects and effects on the ability of other agencies to regulate prices, but in recent electric utility mergers the Securities and Exchange Commission has not conducted substantial antitrust investigations.

Federal Energy Regulatory Commission. Section 203 of the Federal Power Act requires that the Federal Energy Regulatory Commission (the Commission) review electric utility merger applications and reject those that it determines are not consistent with the public interest. Merging parties are required to submit analyses of issues of interest to the Commission, including competitive effects and cost savings. There is then a notice period during which interested parties, or "intervenors," are allowed to submit comments. Based on its review of the application and intervenor comments, the Commission sets specific issues, such as competitive effects and cost savings, for hearings. Commission staff, intervenors, and applicants then undertake extensive discovery of the other parties to obtain information and documents, and Commission staff and intervenors file written testimony. There are then additional rounds of written rebuttal and possibly surrebuttal testimony, as well as depositions. While the matter could be confined to written filings, an administrative law judge typically conducts live hearings before issuing an initial decision. Following intervenor comments on the initial decision, the full Commission issues an order, which may be appealed to both the Commission and the courts. The whole process can take several years. The Commission's review of Northeast Utilities' acquisition of Public Service of New Hampshire took two years from filing (January 1990) to Commission approval after rehearing (January 1992). The merger was completed five months later. However, it was March 1993 before the Commission issued its order on tariffs in connection with transmission conditions imposed on the merger, and it was May 1993 before the last major legal challenge to the merger was removed by a federal appeals court.

Federal Trade Commission and Department of Justice. The Hart-Scott-Rodino Antitrust Improvements Act of 1976 requires that parties to mergers and acquisitions that exceed a certain size notify the Federal Trade Commission and the Antitrust Division before consummating a proposed transaction. Both the acquiring business and the business being acquired must submit information about their respective operations and wait a specified period. During that waiting period, one of the enforcement agencies conducts a preliminary review of the antitrust implications of the proposed transaction.

The federal antitrust agencies have the authority to investigate and challenge electric utility mergers, and they have issued complaints in the case of mergers among regulated natural gas pipeline companies. In the past, however, the antitrust agencies have deferred to the Federal Energy Regulatory Commission to investigate the competitive effects of electric utility mergers. In the case of the proposed merger between Southern California Edison and San Diego Gas and Electric, the Justice Department staff participated in the investigation of competitive effects only as an intervenor in the Commission proceedings.

If an antitrust agency decides that a transaction may present a competitive problem, it can issue a request for additional information commonly called a "second request." At this stage there can be significant delays. The requests can be quite extensive. Compliance involves producing dozens to hundreds of boxes of documents. In addition, second requests routinely compel creation of information, including detailed sales information and analyses of entry costs. Two to three months may be needed to comply with a request.

After review of this additional information, an antitrust agency may choose to challenge the transaction. The Department of Justice can seek a preliminary and permanent injunction in federal court. The Federal Trade Commission has similar powers, though they differ in important details.

State Review

Electric utility mergers must also pass review by the appropriate state regulatory commissions. The nature of the review can vary widely depending on the particular circumstances. Some states have extensive reviews that include competitive effects and cost savings.

The review of the proposed merger of Southern California Edison and San Diego Gas and Electric was very extensive. In that case, the companies filed with the California Public Utilities Commission in December 1988. Four prehearing conferences were held between February 1989 and April 1990. Then the commission held thirteen public hearings in various locations throughout the two utilities' service areas during April and May. From May until August, the administrative law judges presided over sixty-one days of evidentiary hearings in which twenty-one parties participated. Testimony was presented by 116 witnesses, sixty-one of whom made an appearance at the hearings. In February 1991, the administrative law judges recommended against the merger in a 1,253-page decision. After hearing oral arguments, the California commission in May 1991 disapproved the merger because Edison had not demonstrated that consumers would benefit from the merger, because the merger would adversely affect competition in the areas of wholesale transmission and bulk power, and because the merger was not in the public interest. In total, it took two and a half years for the commission to make a decision.

State attorneys general may also review a proposed acquisition. Under the federal antitrust laws, states may seek an injunction in court to enjoin a merger. However, no state attorney general has done so in recent years in connection with an electric utility merger (Coate 1993). Like the Department of Justice, a state attorney general may choose to participate in other reviews rather than going to court. The California attorney general intervened in the state commission review to oppose the proposed California utility merger. Had the state commission approved that merger, the California attorney general could still have challenged the merger in court under state and federal antitrust law.

Private Review

Interested private parties may participate in many different ways, including directly seeking an injunction in federal court. At the federal level, third parties can lobby the federal antitrust agencies to investigate and challenge anticompetitive aspects of a proposed acquisition. Private parties can intervene in dockets before the Federal Energy Regulatory Commission and the Securities and Exchange Commission. Parties may also be able to intervene in dockets before state regulatory commissions. In addition, it may be possible to lobby state attorneys general to become interested in the competitive effects of a merger. In fact, such lobbying is fairly common in mergers and acquisitions in other industries.

If the various government agencies do not prevent a merger, private parties can seek relief directly from federal courts. A plaintiff would likely have two potential avenues for relief. First, a plaintiff could claim that one or more of the review agencies did not fulfill its statutory obligation in reviewing the merger. For example, in 1991 the Holyoke Gas and Electric Department, a municipal utility, filed suit in federal court in opposition to Northeast Utilities' purchase of Public Service of New Hampshire. The Securities and Exchange Commission had initially approved the merger, saying that it would have no anticompetitive effects. Holyoke and others immediately challenged this finding by noting that the Federal Energy Regulatory Commission was discovering major competitive problems. The Securities and Exchange Commission issued a second order contingent upon Federal Energy Regulatory Commission approval of the merger. Holyoke filed suit alleging that the Securities and Exchange Commission had abdicated the antitrust review required by the Public Utility Holding Company Act. Although the court ultimately did not overrule the Securities and Exchange Commission approval, it did state that the Securities and Exchange Commission could not shirk its statutory mandate by simply relying on the Federal Energy Regulatory Commission's concurrent jurisdiction.[2]

Second, a plaintiff could claim that the proposed acquisition would violate Section 7 of the Clayton Act. To bring such an action, the plaintiff would have to have standing under the federal antitrust laws. This standing requirement tends to be more limiting than the standing requirements generally applied by regulatory agencies. The plaintiff must show that it is likely to be harmed by the alleged anticompetitive effect from the merger. Between 1982 and 1992 at least twelve firms sought relief under Section 7 of the Clayton Act, and eight of these firms succeeded (Coate 1993). Given the extensive review by other agencies, however, it is not clear that private challenges to electric utility mergers would enjoy the same degree of success.

CONCLUSION

Competition among producers seeking their own selfish interests is a powerful force to lower costs and prices and to supply consumers with the goods and services they want. Consumers are the ultimate beneficiaries of the competitive process.

The existence of competition cannot be taken for granted. The exercise of market power by monopolists, dominant firms, and cartels denies the benefits of competition to consumers. The antitrust laws are intended to limit the acquisition and exercise of market power. For example, Section 7 of the Clayton Act seeks to prevent mergers that would result in market structures in which the exercise of market power would be likely.

Public utility regulation is another government policy designed to place limits on the exercise of market power. It is a very imperfect substitute for competition, however. This has two important implications. First, even where there is extensive regulation, as in the electric power industry, consumers are likely to benefit from the competition that does exist. The existence of regulated transmission tariffs, open access requirements, and the like should not be an excuse for approval of mergers that reduce competition. Second, where workable competition is likely, it makes sense to avoid public utility regulation.

Changing market conditions and regulations are leading to greater competition in the electric power industry, which appears to be following other industries in its response. Both the commercial airline industry and the natural gas industry experienced a large increase in the number of mergers and acquisitions during and following changes in regulations promoting competition. The electric utility industry has also seen an increase in merger and acquisition activity.

An important concern is whether the antitrust review of electric utility mergers will be carried out appropriately. Many of the mergers in the airline industries were reviewed under special procedures by the Department of Transportation rather than by the federal antitrust agencies. At times, the Transportation Department did not heed the competitive concerns of the Department of Justice. Recent experience indicates that the Federal Energy Regulatory Commission may not subject electric utility mergers to traditional antitrust analysis. The Commission may approve mergers that reduce competition provided the merging parties accept certain "open access" transmission conditions (Frankena and Owen 1993b).

NOTES

1. *Brown Shoe Co. v. United States,* 370 U.S. 294, 320 (1962).

2. *City of Holyoke Gas & Elec. Dep't. v. Securities and Exch. Comm'n.,* 972 F. 2d 358 (D.C. Cir. 1992).

2

Competition and Regulation in the Electric Power Industry

The importance of antitrust analysis of electric utility mergers increased during the 1980s and can be expected to increase in the 1990s. Expanding competition in the sale and transmission of bulk power, and accompanying changes in regulation, create pressures for structural change in the electric power industry. Interest in mergers as well as in divestitures and innovative contractual relationships is growing. The expanding role of competition has increased the potential for mergers to have both anticompetitive and procompetitive effects that deserve evaluation.

In this chapter and the next we evaluate regulations and recent developments in the electric power industry that affect the competitive analysis of mergers. In Chapter 3 we will discuss wholesale power markets and nonutility generation, the Energy Policy Act of 1992, and market-determined pricing for bulk power. Here we begin with a brief review of traditional regulation at the state and federal levels. The privately owned segment of the electric power industry remains subject to pervasive regulation of prices and entry. Nonetheless, at the wholesale level, competition is already significant, and recent regulatory changes can be expected to increase that competition. We believe that there is considerable scope for competition in generation and significant scope for competition in transmission and distribution. Finally, we examine retail competition in two analogous industries: cable television and telephone service.

THE REGULATION OF ELECTRIC UTILITIES

In the United States, investor-owned utilities account for 76 percent of retail distribution of electricity, excluding self-generation. The remainder is accounted for mainly by municipally owned and cooperative distribution systems (Energy Information Administration 1992, Table 1).

Regulation of investor-owned utilities is pervasive but fragmented between state and federal agencies. Under state and municipal regulations, investor-owned utilities normally have exclusive territorial franchises to sell retail electricity. Investor-owned utilities are subject to state regulation of retail prices, and they are required to provide electricity to all customers on a reliable basis. Traditional rate regulation is based on the cost of service, with an allowed rate of return on a base consisting of the depreciated historical cost of assets.

State regulators have considerable powers over investment and operating decisions of utilities. Utilities typically must obtain state approval to construct and site generating and transmission facilities. Under certain conditions, state regulators can disallow costs that otherwise would enter a utility's rate base (for example, capital costs associated with power plants) or that could be passed through to retail rates (for example, costs of purchased power). State commissions also have the power to disapprove mergers of electric utilities.

Under the Federal Power Act of 1935, the Federal Energy Regulatory Commission regulates rates and other conditions for wholesale sales by privately owned utilities, including sales of power and transmission services. Rates at which investor-owned utilities sell power and transmission service to "requirements customers," such as municipally owned and cooperative distribution systems, are regulated based on cost-of-service and rate-of-return principles. Rates for "coordination" transactions between investor-owned utilities for economy energy and short- and medium-term capacity and energy are subject to more flexible regulation, with the result that competition often plays a significant role in price determination.

Under the Energy Policy Act of 1992, the Commission also has the power to regulate rates for sales of bulk power by so-called exempt wholesale generators, and it can order utilities to provide wholesale transmission service to generators of power. It can also order utilities to expand their transmission capacity in order to provide wheeling services, but the states must still approve transmission investments and recovery of their costs in retail rates. Furthermore, under Section 203 of the Federal Power Act, the Commission can disapprove mergers between electric utilities. Mergers must be consistent with the public interest. In determining consistency with the public interest, the Commission investigates, *inter alia*, competitive effects and cost savings.

The Commission also regulates the activities of power pools, and it licenses nonfederal hydroelectric projects. Otherwise, it does not significantly regulate investment in generation and transmission facilities or entry. Of course, where an investment is dependent upon approval of a wholesale power transaction by the Commission, the Commission plays a role in expansion and entry decisions.

The U.S. Nuclear Regulatory Commission must approve licenses for nuclear power plants. Its jurisdiction includes antitrust considerations, and it has made approval of licenses for certain plants conditional on acceptance of wheeling obligations. Wheeling obligations have also been imposed by the courts. The U.S. Supreme Court found that Otter Tail Power had engaged in anticompetitive practices and required Otter Tail to wheel power for municipal utilities as a remedy under the antitrust laws (*Otter Tail Power Co. v. United States,* 410 U.S. 366 (1973)).

Under the Public Utility Holding Company Act of 1935, the Securities and Exchange Commission imposes burdensome regulations on the ownership, organizational structure, financing, transfer pricing, and other aspects of electric and gas holding companies in order to prevent financial and regulatory "evils...connected with public utility holding companies" such as those that occurred during the late 1920s and early 1930s. This act was designed to regulate holding companies that operate in more than one state; those operating in a single state are generally exempt. Until it was amended by the Energy Policy Act of 1992, the Public Utility Holding Company Act substantially limited entry by potential low-cost generators of long-term bulk power. Among other matters, mergers and acquisitions involving electric utility holding companies are subject to approval by the Securities and Exchange Commission.

While regulation of the electric utility industry is pervasive, some aspects of the industry are not regulated extensively. First, publicly owned entities—federal power projects and marketing agencies, municipal utilities, electric cooperatives—generally are not regulated or, in the case of municipal utility investments, are regulated less extensively than private utility investments. Second, the planning of interstate transmission networks is not regulated in a comprehensive manner, although the individual states regulate the intrastate portions of those networks and the Federal Energy Regulatory Commission regulates some of the activities of power pools. Power pools and the North American Electric Reliability Council have acted to coordinate the planning, investment, and operation of interconnected utilities, often across state lines.

Deficiencies of Regulation

Traditional cost-based rate regulation of the type employed by the Federal Energy Regulatory Commission and state utility commissions has well-known deficiencies. First, cost-based regulation reduces a utility's incentives to minimize costs, which are passed along to customers in regulated prices. Cost-based regulation also creates incentives for inefficient behavior. Under certain conditions (for example, where the allowed rate of return exceeds the cost of capital), regulation may induce a utility to engage in wasteful investment.

Second, cost-based regulation results in prices that do not reflect scarcity values, and hence it wastes resources. Prices are usually based on past costs rather than current replacement costs and on long-run average costs rather than relevant marginal costs. One perverse result of cost-based regulation is that prices are increased when demand drops and excess capacity increases. Typically, prices are adjusted to reflect changes in costs only after significant delays. (Such delays, however, increase the incentive for utilities to reduce costs because cost reductions are not immediately passed on to customers in lower prices.) When regulated prices are below marginal economic costs, incentives to invest in capacity are inadequate and consumers are induced to consume too much electricity. When prices are set at the wrong levels, the allocation of resources may be inferior to that which would result from the exercise of market power.

Wholesale pricing of energy (not including capacity) sold by investor-owned utilities is a different case. The Commission's general policy is that energy prices should be based on short-run marginal costs, although various "adders" are permitted to cover assumed costs that cannot be measured. In addition, the Commission allows more flexible pricing for power transactions among investor-owned utilities.

Third, traditional cost-based regulation is expensive and imperfect. Regulators use vast amounts of public and private resources to determine which costs they should allow utilities to include in their prices. In practice, however, regulators cannot prevent utilities from exercising market power (for example, by using cross-subsidies and inflated prices paid to unregulated affiliates to obtain approval for higher retail rates).

The problems associated with cost-of-service regulation could be reduced if regulated prices were replaced by price caps that were not set on the basis of the costs of the particular utility in question. With such regulation, utilities would have a greater incentive to minimize costs.

In addition, utilities could not evade regulation by inflating their costs through cross-subsidies and transactions with unregulated affiliates.

Implications for Competitive Analysis

The deficiencies of regulation have important implications for the competitive analysis of mergers in the electric power industry. First, competition, whether or not it is sufficient to justify elimination of regulation, can reduce the costs of regulation, reduce the exercise of market power, and improve the allocation of resources. Competition should therefore be protected where it exists and fostered where it can be increased.

Second, as the structure of the electric power industry changes, efforts should be made to avoid unnecessarily increasing the burden on regulators. Unless cost-of-service regulation is replaced by an alternative, such as price caps, much can be said for limiting the role of regulated monopoly utilities in unregulated competitive markets for bulk power, and particularly for prohibiting power purchases by investor-owned utilities from unregulated affiliates.

Third, the criterion for deregulating prices should not be the existence of perfectly competitive conditions. Some risk that market power may be exercised from time to time and place to place may be an acceptable price to pay for lower costs and greater efficiency. The U.S. Department of Justice (1984a, 28) has recognized that the costs of regulation make it efficient to deregulate prices in a market even if concentration is higher than would be permitted for a merger under the incipiency standard of Section 7 of the Clayton Act. Furthermore, regulation should not be used to suppress the temporary exercise of market power during a transition from a regime with pervasive regulation to one with greater reliance on market forces. Transition prices serve as important signals to guide the efficient reallocation of resources. In any case, to the extent that regulation causes some prices to be below competitive levels, one should expect those prices to rise when they are deregulated, even absent market power.

THE SCOPE FOR COMPETITION

There has been widespread interest in increasing the role of wholesale competition in determining resource allocation and pricing in the electric power industry. This interest arises because of the superiority of competition to regulated monopoly in reducing costs and prices, and because of the problems utilities have at times faced in getting regulatory approval to raise rates to recover their costs. Many changes in regulation and structure have been proposed (Office of Technology Assessment 1989, Joskow and Schmalensee 1983).[1] The principal objective of these proposals is to increase the role of competition in the generation and sale of bulk power.

Some proposals to increase competition involve changes not only in regulation but in the vertically integrated structure of the privately owned segment of the electric power industry. At a minimum, nonutility generators would own a substantial share of new generating capacity. Under some proposals, the role of investor-owned utilities in wholesale power sales would be reduced, and transmission facilities would become common carriers. In the most radical proposals, common ownership of activities at different vertical stages would end.[2]

Critics of increased competition have pointed to the problems associated with planning and operating an interconnected electrical system in which many disparate entities make investment and operating decisions about generation and transmission. They fear structural changes that would increase competition would raise the cost of electric power and reduce the reliability of the entire electrical system (Gegax and Nowotny 1993). At the very least, others claim, the effects of these structural changes on the cost and reliability of the electrical system are unknown. Because these changes allegedly create huge risks, they argue that reform should proceed very slowly.

The electric power industry consists of three vertically related activities: generation, transmission, and distribution. The scope for increased reliance on competition can be analyzed in two steps. The first is, what is the efficient organization of each of the three activities taken separately? Would the number of competitors be too small for outcomes to be workably competitive? Would any costs involved in increased competition be greater than the alternative costs and problems of regulating firms with market power? As we shall see, there is considerable scope for competition in generation and significant scope for competition in transmission and distribution.

The second step is, would problems of coordination among independent entities entail heavy costs and reliability problems, as the opponents of change contend? Would the problems of coordinating the three vertical levels of the industry under separate ownership notably constrain the scope for competition? If the scope for competition in transmission or distribution is limited, are there substantial economies of vertical integration that constrain the scope for competition in generation? We have seen little evidence of this.

The potential gains from increased competition and reduced regulation in the electric power industry are great. To obtain the full benefits of deregulation and competition, significant structural changes in the industry are needed. The electric power industry in the United States developed largely as a set of isolated geographic monopolies regulated by public bodies. Arbitrary and accidental boundaries, monopoly status, and regulation have distorted the structure of the industry and limited adjustments to changing technology. There is no reason that the geographic extent of local utilities, established many decades ago, should be efficient for today's technologies. Monopolists have too little incentive to minimize costs and face too few penalties when they make incorrect investment decisions.

For all of these reasons, greater competition is likely to create pressures for structural change: mergers, hostile takeovers, divestitures, early plant retirements, and so on. Indeed, the greater these pressures, the greater the justification for introducing more competition. It would be ironic to use the compelling evidence in favor of greater competition as an argument against such change. Where the distortion of investment decisions and other features of industry structure is most severe, prices will be most seriously distorted.

The prospect of a temporary departure of prices from those that would prevail under competition is not a sound basis for opposing change. Temporary local expressions of market power accompanying the loosening of regulation are the very signals that are needed to guide the industry toward a more efficient structure. The opportunity to earn short-term supracompetitive profits can be an effective incentive to make procompetitive investments in restructuring the industry. Overzealous regulation may smother the very signals needed to induce beneficial change.

Competition in Generation

In considering the scope for increased reliance on competition in generation, the first question is how competitive relevant markets would be. In evaluating this issue, it is useful to distinguish between long-term, medium-term, and short-term bulk power.

Whether market-determined prices for delivered bulk power would be competitive depends largely on regulations affecting entry (for long-term power); patterns of control over generating capacity (for medium-term and short-term power); and transmission costs, capacity, access, and pricing (for all three kinds of power). Here we will assume that regulatory barriers to entry by power plants have been removed (except for reasonable regulations designed to take account of environmental effects and other externalities). We also assume that transmission is available on a competitive basis.

For long-term bulk power, the competition is in building, owning, and operating new generating capacity to supply power to wholesale customers under long-term contracts.[3] One reason for contracts to last twenty years or longer is to prevent opportunistic behavior by purchasers of bulk power and by transmission companies, which might otherwise exercise market power once power plant investments have been made. Opportunistic behavior will be less, and reliance of new power plants on shorter-term contracts and spot sales will be greater, if power plants have guaranteed access to transmission at efficient prices as a result of the Federal Energy Regulatory Commission's implementation of the 1992 Energy Policy Act.

Most interest in nonutility generation relates to long-term bulk power markets. The principal issue is whether conditions for entry into the market for developing, owning, and operating power plants are such that there would be numerous potential bidders to satisfy demands for power. The answer depends on economies of scale at the firm level in owning and operating power plants, as well as on economies related to learning that may give an advantage to experienced firms.

In principle, substantial economies of any sort could limit the number of firms that could compete to set up a new power plant. In fact, however, long-term bulk power markets would be highly competitive. Recent experiences with nonutility generators negotiating or bidding to supply new capacity to distribution companies make this clear. The large number of bidders that have participated in competitive auctions to supply long-term bulk power to distribution companies is one indication of the substantial scope for competition in allocating resources and determining prices for new generating capacity. As of 1991, an average of eight megawatts had been bid for each megawatt requested (Elston 1991, 5). Bidders include architect-engineering firms, equipment suppliers, and other companies that design and construct utility-owned power plants, as well as industrial companies with experience as cogenerators. Of course, even with many bidders, competitive problems can remain. Market power can be exercised for complementary transmission services, and generating costs can be affected by regulatory barriers to entry by generators.

Even if there were economies of scale in generation at the firm level, the market presumably would be national. Thus there should be room for numerous competitors. In addition to roughly fifty regulated utilities with affiliates involved in independent power, more than fifty nonutility firms are active in developing, owning, and operating generating plants (Joskow 1991, 69–70). In Joskow's view, entry and exit appear "quite easy."

The Energy Information Administration (1993, ix) recently observed:

> There is concern in the utility industry that the addition of many nonutility generators to the [electric power] system may make it harder to control or even destabilize it. . . . These concerns appear to be unwarranted. As long as certain technical requirements are met by the additional nonutility generating facilities, it should be possible for the electric power system to accommodate them without any decrease in its reliability.

For medium-term power, the competition is among firms to supply bulk power between one and several years in the future. In the near future, the medium-term bulk power market will largely involve sales from unplanned excess generating capacity owned or controlled under long-term contracts by distribution companies (including investor-owned utilities), as well as sales from public power authorities. Nonutility generators whose initial contracts for twenty to thirty years have expired will also be in the market. Given adequate transmission access, new nonutility generating capacity might be active in this market if such generators are efficient bearers of relevant risks.

For short-term power, competition is among firms to supply bulk power to wholesale customers under contracts lasting anywhere from an hour to a number of months. The suppliers to the short-term bulk power market typically are distribution companies (including investor-owned utilities) that are selling power, particularly economy energy, associated with generating capacity that they either own or have under contract.

Competition in medium- and short-term power will depend not only on existing capacity but also on new capacity. Hence it will depend on the economies of scale for new generating firms. Partly because of changes in technology, these economies of scale are likely to be relatively limited. Furthermore, large power plants can be jointly owned by competing firms.

In considering the scope for competition to replace regulation in wholesale power supply, we have been assuming that transmission service would be available from existing facilities on competitive terms. Nonetheless, it is important to consider the cost of transmission and its availability, since these affect the geographic scope of markets and thus concentration, particularly for medium-term and short-term power.

The costs of transmission are sufficiently low that bulk power markets generally would be competitive, given existing control of generating facilities, provided transmission is available on competitive terms. Most of the existing problems of market power in wholesale power are related to control over transmission rather than control over generation. Of course, there are likely to be exceptions, particularly in the short term as a result of limits on existing transmission capacity.[4] Sales of bulk power to small, remote distribution systems may also be an exception.

In long-run decisions regarding the location of new generation facilities, the capital costs of generation and transmission must be considered. There are many situations where it is efficient to transmit power for hundreds of miles, and some situations where it is efficient to transmit power for a thousand miles. Particularly long distances are efficient when generating costs depend heavily on location (for example, because of availability of hydroelectric sites, emissions control policies, or economies of scope for production of power during both winter and summer or during both peak and off-peak hours). Transmission costs per kilowatt-hour are also lower, and hence the distance over which transmission is efficient is greater, for large power flows because of economies of scale in transmission relating to capital costs and electrical losses (for example, use of higher voltages and direct current).

Consider what is happening in the West:

> Utilities with coal-fired capacity in Utah, Wyoming, Arizona, and New Mexico sell power to California to back out oil and gas. Coal-fired electric power is also sold to the Northwest to supplement hydropower during dry spells and during winter peak periods. The Pacific Northwest sells hydropower to California utilities and other Southwestern States when water conditions permit and during summer peak periods. (Office of Technology Assessment 1989, 195)

In addition, utilities in the Southwest import power from Mexico. Very long transmission lines have been built to transmit remote low-cost Northwest and Canadian hydropower, power from mine-mouth lignite plants in North Dakota, and power from the bituminous coal-fired Four Corners and Intermountain projects in New Mexico and Utah (Federal Energy Regulatory Commission 1989a, 56). Southern California's ban on coal-fired power plants may influence the latter.

In the short run (when capital costs of generation and transmission are sunk), opportunities for efficient long-distance transmission are likely to be great. There are two reasons for this. First, most long-run transmission costs are accounted for by capital costs. In the short run, given available transmission capacity, the principal cost of transmitting energy is electrical losses. Thus, in the short run, transmission costs are relatively low. For a typical 100-mile, 345-kilovolt line, losses are about 1 to 3 percent of the power transferred (Federal Energy Regulatory Commission 1989a, 60-61). Line losses per mile per unit of energy transmitted increase with distance and energy flow on a line of given capacity and decrease with voltage. Other short-run transmission costs relate to reactive power supply and voltage control.

Bulk power transactions involve not only energy but also capacity transactions. In some capacity transactions, the purchaser is buying an option to obtain energy from a particular source (for example, as backup in case of a power plant or transmission line outage or an unexpected increase in demand). In nonfirm energy transactions, the purchaser buys only energy on an interruptible basis, and does not buy capacity rights. In power transactions involving capacity where the probability of actually transmitting the associated energy is low, transmission costs are even less important.

Second, the short-run variable costs, mainly fuel costs, associated with different generating technologies vary more than the long-run costs, and they are substantial compared with short-run transmission costs. Thus, in the short run, low-cost generating capacity can compete in sales over relatively large geographic areas.

In summary, no inherent structural features relating to concentration and entry prevent the generation and supply of bulk power from being competitive in most areas. Of course, even with open transmission access at competitive prices, there could be market power in some markets for bulk power, particularly medium- and short-term power. There may be constraints on transmission capacity or concentrated ownership of generating capacity. As generation becomes more competitive, one problem will worsen: opportunities will increase for evasion of cost-based regulation of retail prices. Where investor-owned utilities with regulated monopolies have affiliates in competitive bulk power markets, competition may be foreclosed.

Characteristics of Transmission Systems

Three important characteristics of transmission systems influence the efficient organization of markets. First, the transmission of power between two points is subject to economies of scale. One high-capacity, high-voltage line has lower unit capital costs and lower unit electrical losses than two low-capacity, low-voltage lines (Federal Energy Regulatory Commission 1989a, 60, 212–13).

Second, the operation of transmission and generating facilities as an interconnected system has substantial benefits. Investment, operating, and maintenance decisions can be coordinated. There are economies of scale in the supply of reserve transmission and generating capacity to achieve a given level of reliability or protection against unanticipated electricity demands and equipment failures. In addition, wholesale power transactions among facilities on a transmission network reduce capital and operating costs.

Third, in an interconnected electric power system, capital and operating costs for the owner of one part of the system depend on capacity and use elsewhere on the system. If different parts of the system are under independent ownership, there are external economies and diseconomies associated with investments in and use of transmission capacity. External diseconomies arise when an investment or operating decision imposes costs on a third party, other than effects involving prices of outputs and inputs. Use of an interconnected electric system by parties that are linked by a contract path may make entities that are not parties to the transaction worse off in various ways: the third parties may have to forgo profitable purchases or sales of power, may have to reduce output from low-cost generating plants while increasing output from high-cost generating plants, or may have to expand the capacity of their transmission facilities.

Consider a transmission system that connects the vertices of a triangle, points A, B, and C. Power is transmitted from A to B by increasing generation at A and increasing electrical load at B. As a matter of physics, electrons will flow along all paths connecting A and B, with the allocation of energy flow inversely related to impedance on those paths. Impedance depends on distance, voltage, and energy flow on a line. The energy that flows from A to C to B is known as parallel path flow. Because of parallel path flow, the capacity of the transmission system (based on thermal and stability constraints) to transmit power along a contract path from B to C, and the cost of transmission in terms of electrical losses, will depend on capacity and use of transmission facilities on the contract path between A and B (Simons et al. 1993). Similarly, the overall capacity of a network to transfer energy from one set of nodes to another across any imaginary line depends on the pattern of generation and loads at various nodes on the network (Hogan 1992).

Because of these characteristics of transmission systems, the costs of transmission and generation would be minimized by the following actions:

- Investment in individual transmission and generating facilities that exploit economies of scale.
- Integration of transmission facilities connecting generating and load centers in a given geographic area.
- Interconnection of transmission systems over geographic areas with substantial amounts of generating capacity and loads.
- Coordination of investment in and operation of transmission and generating facilities in an interconnected system. One way to achieve efficient

operation of a given system would be through a power pool in which the generating facilities used would be determined centrally to minimize costs, given the transmission facilities.

Putting this together, one might think that a regional or larger transmission grid (perhaps along with generating facilities) is a natural monopoly that should be owned and operated by a regulated monopolist. In principle, such a monopolist could be organized as a regulated tight power pool that allocates resources efficiently rather than maximizing profits through monopoly pricing and foreclosure of competition (Hogan 1992). Opponents of increased competition have used the preceding characteristics of transmission systems to support the status quo.

There are problems with the line of argument that the external economies and diseconomies on an electric network prevent competition in transmission and generation. Suppose, for the sake of argument, that a given set of transmission facilities is a natural monopoly. Competition in the transmission services produced using those facilities still could play a significant role in allocating resources and determining prices, as we discuss below. This is particularly true in light of the costs involved in regulating a monopoly. Also, there are likely to be market mechanisms that would permit the industry to take account of externalities and economies of scale without monopoly. For example, the Federal Energy Regulatory Commission is planning to consider how transmission pricing can account for the effects on third parties of parallel path flow. The broader point is that there are several solutions to problems of externalities besides common ownership, most notably the establishment of more complete property rights that permit efficient contracting.

Competition in Transmission

The fact that transmission networks have certain natural monopoly characteristics does not imply that transmission systems must be operated by monopolists. In the electric power industry and elsewhere in the economy, facilities that are subject to economies of scale are operated as joint ventures in which the partners can cooperate in certain activities (such as construction and operation of the facility), make unilateral decisions in other areas (such as capacity, output, and pricing), and compete in marketing and sales (Smith 1987, 1988, 1991; McCabe et al. 1991; Houston 1991). Such "competitive rules" joint ventures among firms that remain competitors exist not only in electric power transmission but also in electric power generation, oil and natural gas pipelines (Alger and Braman 1993), automatic teller machines, credit cards, airline computer reservation systems (Wildman and Guerin-Calvert 1991), newspapers (Reynolds 1990), and aluminum can body stock (*United States v. Alcan Aluminium Ltd., et al.* 1985-1 Trade Cas. (CCH) ¶66,427, ¶66,428 (W. D. Ky. 1985)). They have also been proposed for deepwater ports (U.S. Department of Justice 1976).

Some transmission lines are owned by several different parties, each of which has rights to a certain amount of capacity. Similarly, the capacity of some lines has been allocated among several different parties (investor-owned utilities and local distribution companies) under long-term contracts. In at least some cases, such as the integrated transmission system operated by Georgia Power and the Municipal Electric Authority of Georgia (La Bella 1989) and the Pacific Interties connecting the Pacific Northwest and California, the holders of transmission rights can compete in supplying transmission service. Even where wheeling is limited by contract or by the incentives created by regulation, the holders of transmission rights compete in supplying transmission service by buying power in one location, transmitting it, and reselling it in another location.

In other cases, different parties own or have transmission rights on different transmission lines that are part of an interconnected transmission system, and they use these rights to compete with each other in supplying transmission service. For example, both Southern California Edison and San Diego Gas and Electric have transmission rights that they can use to transmit bulk power from coal and nuclear generating plants in the Southwest (Arizona and New Mexico) to buyers such as Pacific Gas and Electric in northern California. In evaluating the proposed merger of Southern California Edison and San Diego Gas and Electric, the Department of Justice, the Federal Energy Regulatory Commission administrative law judge, and the California Public Utilities Commission all determined that the merger would eliminate significant competition between these two utilities in short-term transmission from the Southwest to northern California. More recently, the Federal Energy Regulatory Commission has determined that there was significant competition in transmission between other investor-owned utilities that have merged, such as Northeast Utilities and Public Service of New Hampshire.

A report by federal regulators has speculated that market power would be limited by competition among the owners of a transmission joint venture and by competition on the part of purchasers of transmission service that are allowed to resell their transmission rights (Federal Energy Regulatory Commission 1989a, 73, 109–12, 119, 144, 164, 168–69, 174, 178–79). This report also claims that similar "head-to-head" competition among owners of different transmission lines that are part of a single grid would be "inherently infeasible" because of parallel path flows. In fact, the latter type of competition is significant (for example, between Southern California Edison and San Diego Gas and Electric).

Although parallel path flows are not yet handled with pricing mechanisms, they are handled in some cases by agreements that allocate the total transmission capacity in a corridor among the owners of transmission facilities. Nonetheless, Stalon, who was responsible for the Commission's recent report on transmission (Federal Energy Regulatory Commission 1989a), maintains that "attempts to create efficiency-increasing competition between TOUs [transmission-owning utilities] that own parts of interconnected networks are almost certainly doomed to fail. While one can imagine a regulatory and pricing system sufficiently sophisticated to create efficient competition, I find it difficult to imagine such results from any regulatory system likely to evolve in the American political system" (Stalon 1993, 34).

In many competitive situations, owners of transmission rights do not have market power, and thus there is no reason for prices to be regulated. Furthermore, given the costs and inefficiencies inherent in regulation, it makes sense to tolerate some risk that market power will be exercised rather than imposing or retaining regulation. Also, the ability to exercise temporary market power as a result of deregulation is likely to prompt firms to make procompetitive investments, and therefore it should not be the basis for rejecting deregulation.

Competition, however, will not be adequate to permit complete deregulation of transmission. As a practical matter, the ownership structure of transmission makes it likely that the exercise of market power in transmission can be limited in many situations only by significant regulation of transmission pricing, access, and expansion, along with enforcement of the antitrust laws. For example, investor-owned utilities have considerable market power over transmission for municipal and cooperative distribution systems and independent generators within their own control areas. Investor-owned utilities may also have market power because of limited alternatives to wheeling power across their control areas.

Given the costs and imperfections of regulation, however, competition can play a useful role in limiting the exercise of market power in transmission. This will remain true even with mandatory transmission access and open access transmission tariffs.

Transactions Costs of Coordination

The Office of Technology Assessment of the U.S. Congress considered the main scenarios that are being proposed for regulatory and structural changes at the wholesale level in the electric power industry, including increased competition in generation and transmission, separate ownership of transmission and generation, and mandatory transmission access. "Concerns that the bulk power system (generation and transmission) is inherently incompatible with competition do not appear to be well founded," it concluded.

> Problems and issues will arise with widespread competition, but they will be much less technical than political and institutional. . . . The greatest challenge to increasing competition in generation and expanding transmission access is maintaining the high degree of coordinated planning and operation among bulk power system components. . . . The key to coordination will be in defining workable institutional arrangements among participants in the power system. Some new physical facilities and improved analytical capabilities may be required, but all these functions can be provided with familiar technology. (Office of Technology Assessment 1989, viii, 15)

In order for an integrated electrical system to operate reliably and efficiently, close coordination of investment and operating decisions for generating and transmission facilities is required. In evaluating the scope for competition, however, the issue is not one of technical feasibility but of economics, of benefits of increased competition compared with increased costs of coordination through contracts and other means.

We have seen no evidence that the costs associated with negotiations and contracts between independent entities, as well as forgone benefits of coordination, exceed the gains in efficiency that are derived from greater competition.[5] On such issues, the burden of proof should lie with those who favor regulation. In fact, because of the fragmented ownership of generation and transmission, as well as the development of extensive wholesale power markets, coordination is already being achieved on an extensive scale by negotiation and contract, ranging from informal agreements to formal power pool activities. For example, economies of scale in transmission and generation are regularly exploited by joint investments.[6] Interdependencies in transmission capacity are handled by agreements that create and assign property rights to various amounts of transmission to different parties (for example, to the owners of different lines connecting points A and B).

Various large electric utilities have sought approval for mergers in recent years. In many cases, merging party claims of substantial efficiency gains relating to consolidation of ownership over generation or transmission, or integration of transmission and generation, have been rejected or significantly reduced by the regulatory authorities. An important reason has been the recognition that many benefits of coordination among proximate investor-owned utilities can be or already have been realized by negotiation and contract, including the activities of power pools.[7]

In addition, regulatory changes may reduce incentives for vertical integration. For example, mandatory transmission access should reduce one motive that has been suggested for vertical integration: to prevent the exploitation of parties after they have made nonrecoverable investments in generation or distribution, in situations where they could not adequately protect themselves through contracts.

Joskow and Schmalensee (1983, 76, 147, 193–95) have speculated that increased wholesale competition among utilities is likely to make power pools less stable and that it may impede their efficient operation (for example, by making pool members less willing to reveal the cost information used for central dispatch and by making it more difficult to agree on measures to minimize long-run costs, including unit commitment, maintenance scheduling, and capacity planning). The Office of Technology Assessment (1989, 16, 18) is skeptical of this argument.

Distortions in Wholesale Power Markets

The efficiency gains from increased competition in the supply of bulk power depend on the extent to which the prices of bulk power reflect marginal costs. Prices may deviate from marginal costs as a result of externalities relating to parallel path flow in an interconnected transmission network, and as a result of remaining price regulation. The Federal Energy Regulatory Commission's regulation of transmission pricing does not take into account the effects of parallel path flow, but some of these externalities are limited by contracts that allocate capacity. If investor-owned utilities compete to sell bulk power to a municipal distribution system based on embedded-cost rates, the utility with the lower marginal costs could have the higher prices, and thus lose the sale simply because it had newer facilities or greater excess capacity. A municipal utility that is a requirements customer of an investor-owned utility typically buys power from that utility at a rate based on the seller's average embedded cost, which may be below the seller's marginal cost. If the municipal utility can buy more power than it needs for its retail customers, it may profitably resell bulk power at a price that is above its purchase price but below the marginal cost of generation. In that case, higher cost generation may displace lower cost generation.

Similarly, greater reliance on competition in resource allocation could lead to an inefficient expansion of publicly owned utilities. Those utilities receive subsidies that may enable them to undersell privately owned utilities with lower social costs (Joskow and Schmalensee 1983, 105–7; Kleit and Stroup 1987).

Clearly, there are efficiency arguments for taking account of parallel path flows in transmission pricing and for removing the distortions in prices caused by regulation and subsidies. Distortions in wholesale power markets, however, do not justify limiting reliance on competition. Instead, the distortions can be reduced or eliminated through establishment of property rights and regulatory reform.

Competition in Distribution

The conventional wisdom is that electricity for a given geographic area can be distributed by a single company at lower cost than by two or more companies with separate equipment.[8] In other words, the retail distribution of electricity is considered a natural monopoly. While this makes good sense as a matter of engineering, it does not allow for the possibility that the benefits of competition in reducing costs and prices could outweigh the engineering inefficiencies of operating duplicate systems.[9]

Moreover, competition can be introduced into the determination of prices and resource allocation for a natural monopoly without duplicating distribution systems. Under franchise bidding schemes, competing firms bid for the right to operate a monopoly enterprise. The bidder that offers to sell output at the lowest unit price, or at the best combination of price, quality, and other nonprice characteristics, wins the bid. Franchise bidding has commonly been used for cable television systems.

Under certain circumstances, franchise bidding can be used to prevent the exercise of market power and to avoid the costs of traditional regulation (Demsetz 1968). Critics have argued, however, that franchise bidding may not prevent the exercise of market power in some cases (for example, where the incumbent has great advantages over other potential bidders at franchise renewal time). Critics also have claimed that franchise bidding may not avoid the problems of traditional regulation because of difficulties, at the time of the bidding, in specifying how prices will be adjusted to take account of future events (Williamson 1976).

Net gains from competition in distribution might be possible if monopoly local distribution companies were required to provide large retail customers with wheeling services. Investor-owned utilities already provide some of their wholesale requirements customers (such as municipal distribution companies located within their control areas) with wheeling services for bulk power as well as dispatch and coordination services, and they will provide more services under the Energy Policy Act of 1992. Provision of wheeling services to large industrial users of power would be similar.

Large industrial users are currently lobbying for retail wheeling, which would give them access to competing sources of power. The Energy Policy Act gives the Federal Energy Regulatory Commission the power to order wholesale wheeling but not retail wheeling, and the act precludes "exempt wholesale generators" from making retail sales. Also, the Commission has not imposed retail wheeling requirements as a *quid pro quo* for approval of mergers and market-based pricing. There is considerable controversy concerning whether the states have the authority to order retail wheeling. One political obstacle to retail wheeling is that it could cause utilities to recover more of their costs from other retail customers.

Another issue is whether there are economies of vertical integration between distribution companies, on the one hand, and generation and transmission facilities, on the other. Most distribution companies, other than some municipal and cooperative distribution companies, are vertically integrated into transmission and generation, but this is not necessarily an efficient pattern. There are strong regulatory incentives for vertical integration. Until its amendment in 1992, the Public Utility Holding Company Act of 1935 largely excluded generators other than the local distribution utility and qualifying facilities.

ANALOGOUS INDUSTRIES

Cable television service and telephone service have certain similarities to local distribution of electricity. Each uses a network of wires to deliver services to homes and businesses. Frequently, all three share poles and underground conduit space. Each connects local customers to regional or national networks. All three have been regarded as natural monopolies. By reviewing local competition in these similar industries, we hope to shed light on the prospects for competition in local distribution of electric power.

Cable Television

Cable television is the youngest of these industries, dating from the period following World War II. Originally conceived to improve local television reception in areas shadowed by high terrain, cable grew into a multichannel local distribution system carrying dozens of local and national television signals, most of them not available over the air (Owen and Wildman 1992, chap. 6). The prospects are that cable will provide the equivalent of hundreds of television channels.

Ownership of the approximately 11,000 local cable television systems in the United States is rather unconcentrated. The largest multiple system owner, Tele-Communications, Inc., and its associated company, Liberty Media, have minority or majority ownership interests in systems serving 23.7 percent of all subscribers. Many multiple system owners have interests in one or more of the cable programming services that are delivered by satellite to cable systems.

In 1984 Congress eliminated most local, state, and federal regulation of cable television. Between 1984 and 1992, the number of subscribers and the number of channels offered increased dramatically. Although real prices per channel did not increase, nominal rates per subscriber per month did. This trend met with protests that led to the Cable Television Consumer Protection and Competition Act of 1992. The 1992 act imposed a detailed federal and local rate regulation scheme on cable systems not subject to "effective competition." Under this act, "effective competition" for cable systems exists when one of the following three criteria is met: (1) there is head-to-head competition between a local cable system and a "multichannel video" provider, where the competitor serves at least 15 percent of the households in the area, (2) the local cable system serves less than 30 percent of the households in its area, or (3) the cable system competes with another cable system owned by the local franchising authority. Local authorities are directed to ensure that the rates under their regulatory jurisdiction are as close as possible to the rates that would prevail if the system were subject to effective competition.

The most interesting of the conditions defining effective competition involves competition with a multichannel video provider. Although cable systems obviously compete with over-the-air broadcast signals and with videotape rental outlets, these competitors apparently do not qualify as competitors under the 1992 act. Today the principal example of multichannel competition is a cable "overbuild." There are several dozen communities in the United States (the exact number is unknown) where cable systems compete head to head, although not necessarily throughout their territories.

Cable overbuild competition is frequently motivated by "green mail" possibilities. That is, an entrant threatens to enter or does enter with the objective of being bought out by the incumbent. In these cases there is unlikely to be long-term competition. However, some overbuild systems, such as those in Allentown, Pennsylvania, have remained in business for many years.

Owen and Greenhalgh (1986) suggest that economies of scale are not so great as to make overbuild competition harmful to consumers, depending on the pricing behavior of the rivals. They report that in 1980 head-to-head competition between cable systems existed in about two dozen locations in the United States. Assuming that firms minimize their costs, they estimate that costs would be 14 percent lower for a single system than for two competing systems each of which has half the customers in a given area. They suggest that this is not so substantial as to rule out the possibility of effective competition that would reduce prices to consumers.

More important than overbuild competition in coming years will be competition offered by multichannel, multipoint distribution service and direct broadcast satellite service. Multichannel, multipoint distribution service is a wireless, microwave technology for local distribution of multiple channels to rooftop antennas. A number of such systems have been licensed by the Federal Communications Commission, and in some areas they are beginning to achieve significant penetration. Direct broadcast satellite systems, which use satellites to send multiple channels to relatively small rooftop antennas, are now operating in several parts of the world. The Federal Communications Commission has granted direct broadcast satellite licenses to a number of U.S. entities, and one high-powered direct broadcast satellite has been launched.

Cable systems may also begin to face competition from local telephone companies. Several telephone companies, notably Bell Atlantic, have announced plans to develop and offer some form of video-to-the-home service. While local telephone companies are permitted to transmit video programming as regulated common carriers, they are prohibited from packaging and marketing video programming in their telephone service areas (47 U.S.C. §533(b)). This prohibition prevents likely anticompetitive behavior on the part of local telephone companies in cable television (Owen, 1993a, 1993b).

The technology that permits video services to be offered on the hither-to narrow-band telephone system is called digital compression. Digital compression techniques use computer algorithms to squeeze information into relatively narrow "pipes." Thus digital compression permits cable systems with coaxial cables to offer hundreds of channels on wires that formerly could carry only dozens of channels. The same technology can be used to send a smaller number of video signals over the twisted copper pair of wires that telephone companies have used to deliver voice messages. By building high-capacity fiber optic cables to central distribution points in each neighborhood, telephone companies may be able to offer services equivalent to those of coaxial cable plants with digital compression. The economic viability of a telephone-based system, however, has been questioned (Johnson and Reed 1990). Some cable companies, meanwhile, are exploring the possibility of offering telephone service (and other two-way switched services) on their plants, particularly interconnections with long-distance carriers and national data networks.

The extent of vertical integration of cable systems is generally much less than that of electric utilities. Although multiple-system operators frequently own equity interests in cable satellite network services, these investments were generally made in the early days of cable to ensure an adequate supply of programming that would attract new subscribers. For example, a number of multiple-system operators invested in Turner Broadcasting, the operator of Cable News Network and other noteworthy cable network services, during a period when Turner Broadcasting was experiencing serious financial difficulties. No cable system today owns interests in a substantial fraction of the services that it carries. Nevertheless, policy concerns over vertical integration by cable systems continue, and the 1992 Cable Act contains provisions requiring the Federal Communications Commission to address these concerns.

Cable systems face more local competition than do electric distribution systems, and vertical integration by cable systems into competitive upstream markets (for programming services) is far less than vertical integration of electric distribution companies into competitive upstream markets (for generation). Nonetheless, lawmakers are much more concerned about vertical integration in cable television than in electric utilities.

Telephone Companies

Before about 1970, telephone service in the United States was provided by the monolithic Bell System, a tightly integrated and centrally controlled organization encompassing local distribution, national switching networks, and equipment manufacturing and research.[10] (An analogous electric utility company would be one that integrated, physically and financially, about 90 percent of all regional utilities and that owned all the electric equipment manufacturing plants in the country.)

Nascent competition from equipment suppliers and from long-distance networks began to raise questions about the Bell System's monopolies in the 1960s. By the early 1970s, both the Federal Communications Commission and the Antitrust Division of the U.S. Department of Justice were actively attempting to widen the scope of competition in telephony. The Bell System's unflagging resistance, both in regulatory forums and in the marketplace, led to the antitrust lawsuit that ultimately dissolved the Bell System.

That competition might be superior to monopoly was most obvious when it came to telephone equipment. Telephone regulators had long feared that excessive prices paid to Bell's nonregulated equipment supplier, Western Electric, exported profits upstream and thus evaded regulatory constraints on local and long-distance telephone rates. As might be the case in the electric power business if electric utilities had local monopolies on the sale of electrical appliances, the Bell System argued that competitive provision of customer equipment would lead to safety problems and physical degradation of the network services provided to other customers. Such risks are, of course, present in both networks, but monopoly and vertical integration are not necessary to deal with them. The Federal Communications Commission began to require that the Bell System deal fairly with competing equipment suppliers in the 1970s, and by the time of the dissolution Bell's share of customer equipment sales had declined significantly.

The problem with long-distance service was more complex. Just as local electric power grids offer interconnection to local as well as distant sources of electric power, local telephone lines provide access to other local telephones and to long-distance networks. Firms seeking to compete with the Bell System in offering long-distance services had to interconnect with Bell's local switching machines to gain access to the local telephone distribution network. This the Bell System at first refused to do, and later refused to do except under onerous conditions.

One reason why the Bell System refused to interconnect with competitors was its belief that a system of pricing subsidies was attracting unwarranted entry. Cross-subsidies in the telephone system involve complex methodological issues because much of the costs are shared and fixed, not variable. Essentially arbitrary rules regarding allocations of shared and fixed costs determine whether one service is subsidizing another. Nevertheless, the conventional wisdom was (and still is) that long-distance service "subsidized" local service and that, within long-distance and local service, business customers subsidized residential users. Thus the Bell System saw entry by firms such as MCI as motivated initially by a desire to "cream skim" the artificially high-priced long-distance service offered to business customers.

Another reason for the Bell System's resistance to competition, both in the provision of equipment and in long-distance services, was the notion that the national integrated telephone network internalized what would in a decentralized market system be important externalities. No longer would a single firm be responsible for planning and implementing the design and standards for the network. Equipment manufacturers and purchasers would fail to take account of the effects of their design decisions on other users of the network. Experience has proven that the market is remarkably ingenious in dealing with these problems. Now that we have decentralized ownership of the still physically integrated telephone network, there is no evidence that significant network externalities impair the performance of the industry.

When the Bell System was dissolved in 1984, long-distance services and equipment manufacture were separated from local distribution companies. The new AT&T provided only competitive services. The seven Bell Regional Holding Companies were to offer monopoly local service. The subsidy problem was resolved by requiring all long-distance carriers to pay for access to local Bell switches, and by imposing a gradually increasing tax on local customers, the effect of which was to reduce the subsidy from long- distance to local service as measured by the Federal Communications Commission.

The consent decree that led to the dissolution of the Bell System contemplated a long-term "quarantine" on the business activities of the regional holding companies (Noll and Owen 1988). They were forbidden to make or sell telephone equipment, and they were forbidden to provide long-distance service. These restrictions came under immediate attack by the holding companies both in court and in Congress. Today, except for long-distance service, the Bell companies can offer essentially the same services and equipment that were open to the old Bell system.

One basis for the change in policy toward vertical integration by the local telephone monopolies is the gradual diminution of their local monopoly power. Competition is being provided today, in varying but still small degrees, by customer-owned bypass technologies, fiber optic rings and teleports, cellular telephones, and shared tenant services. Large business customers can put satellite antennas on their rooftops or run cables directly to long-distance carriers, thus bypassing taxes and the high transport fees on long-distance service by local telephone companies. In a number of cities, local competitors are offering high-speed data service using fiber optic rings. These services provide both local data interconnections and access to long-distance services to business users. In each city, local cellular telephone service is provided by the local telephone monopolist and by an outside competitor (often a telephone company based elsewhere). As the capacity of cellular systems increases and prices fall, these services will become better substitutes for ordinary telephone service. Finally, many property owners of large, multitenant buildings offer central electronic switchboard services that reduce the revenues of local telephone companies by aggregating the demand for central office lines. These shared tenant services may require that only twenty lines run to the central office, where previously each one of fifty tenants might have had a direct line. In addition, some shared tenant systems connect directly to long-distance services, bypassing the local telephone company.

The effect of these growing sources of local telephone competition is to put downward pressure on the subsidy to residential services. As prices for residential telephone service rise toward cost, competitive entry may take place there as well. Radio-based telephone services, such as cellular, may be a less costly way to provide telephone service to households some distance from local switches. In addition, cable television systems may begin to compete in offering switched two-way local communications services, both voice and video.

In sum, the cataclysmic events that took place in the telephone business in the 1980s were motivated by the same concerns that arise in the analysis of electric utility mergers: vertical integration, foreclosure of competition, and evasion of regulatory constraints. Continuing debate surrounding the activities of local telephone companies is based on these concerns.

CONCLUSION

The electric power industry in the United States has been closely regulated and heavily integrated for many years. It is doubtful that its present vertical, horizontal, or geographic integration is efficient. Moreover, regulation continues to introduce distortions, impairing the efficiency of resource allocation and creating incentives and opportunities for utilities to engage in anticompetitive behavior.

There is already significant competition in power generation as well as, to a lesser extent, transmission, and there is increasing acceptance by policy makers of a role for competition. The arguments against competition, based mainly on the existence of economies of scale, vertical integration, and network externalities, do not appear to be compelling.

Regulatory reform and increased competition present opportunities to create more efficient market structures, some of which can most readily be achieved through mergers. Other mergers will be proposed that seek to reduce competition or to evade regulatory constraints.

The economic and technical characteristics that make the electric power business interesting—scale economies, vertical integration, network externalities, and regulation—are paralleled in other industries such as cable television service and telephone service. Experience in these industries suggests that competition and deregulation are both feasible and attractive from the point of view of economic efficiency and consumer welfare.

NOTES

1. For discussions of the restructuring of the electric power industries in Great Britain and New Zealand to increase competition in the supply of wholesale power, see Green and Newbery 1992 and Braman 1992, respectively. For an update on New Zealand, see *Electric Utility Week*, June 7, 1993, 15.

2. For arguments against prohibiting electric utilities from owning new generation facilities, see Joskow 1992, 31–32.

3. Joskow and Schmalensee (1983, chaps. 9–11) argue that increased reliance on competition in the electric power industry would lead to extensive vertical integration through complex long-term contracts among participants in the generation, transmission, and distribution of power.

4. Schmalensee and Golub (1984) found that in 1978 ownership of generating facilities in many areas was highly concentrated (that is, the Herfindahl-Hirschman Index, explained in Chapter 5 below, exceeded 1,800). Their computations assumed that generators were not potential competitors if they were not (1) among the nearest set of generating facilities with total capacity equal to four times an area's actual load or (2) located within 200 miles. These geographic areas appear small in the 1990s, assuming transmission rights and capacity are not constraints.

5. For an argument that increased competition in generation, and specifically increased reliance on independent power producers, would not threaten reliability, see National Independent Energy Producers 1991.

6. For a long list of jointly owned generating plants, see Energy Information Administration 1991, app. C. Maine Electric Power and Vermont Electric Power are jointly owned transmission companies (Houston 1991).

7. Kaserman and Mayo (1991) studied 1981 data for seventy-four firms engaged in distribution and/or generation. They concluded that there are significant economies of vertical integration between distribution and generation. Distribution is aggregated with transmission, with output measured by sales to final customers; an alternative specification that included resold power as an additional output variable reportedly produced results that were

not materially different. One difficulty with studies that use observations of the industry when it was heavily regulated is that the results could be a product of regulation.

8. Up to a point, there may also be economies of scale associated with the total size of the distribution system. Joskow and Schmalensee (1983, 59) speculate that "these economies are probably exhausted by relatively small cities."

9. Smith (1987) suggests introducing competition into distribution by eliminating restrictions on entry or operating distribution as a cost center jointly owned by competing marketers of retail power.

10. For a brief survey of the early history of telephony and the events leading to the dissolution of AT&T, see Noll and Owen 1988, 1994.

3

Recent Developments in Wholesale Power Markets

Antitrust analysis of electric utility mergers is important because of the benefits of competition in allocating resources and determining prices. The role of competition in the supply of bulk power is substantial, and recent developments in the industry are likely to increase that role.

We begin this chapter with a discussion of the increasing volume of transactions in markets for bulk power and the growth of nonutility generation. We then examine the Energy Policy Act of 1992, which eliminated entry barriers into the generation and supply of bulk power that were imposed by the Public Utility Holding Company Act of 1935. Next we describe how competition affects prices for bulk power and transmission in spite of regulation. Finally, we address transmission access and the Federal Energy Regulatory Commission's new authority to order utilities to provide transmission service.

WHOLESALE POWER MARKETS

Since the 1960s wholesale power markets have developed along with improvements in transmission, generation-control and coordination technology, expansion of high voltage transmission capacity, and increasing interconnections and power pools among utilities. These developments have allowed utilities to reduce the costs of electric power. In some cases, such as the New England Power Pool, generating facilities owned by different parties are controlled by a central dispatch facility, which adjusts the output of the various power plants to minimize costs.

Sales and exchanges of bulk power—capacity or energy—among investor-owned utilities permit utilities to economize on additions to capacity, for example, where utilities have complementary time-of-day or seasonal power demands, or where some utilities have excess capacity. Bulk power transactions also facilitate achievement of economies of scale in generation and permit greater use of generating capacity with low fuel and other variable costs. Nuclear and hydroelectric generation have lower variable costs for energy than coal generation, which typically has lower variable costs than oil and gas generation.

Power purchases by investor-owned utilities from other utilities increased almost continuously from around 224,000 gigawatt-hours in 1975 to 479,000 gigawatt-hours in 1991.[1] Annual interchanges (exchanges) among investor-owned utilities during this period ranged between 154,000 gigawatt-hours (in 1991) and 412,000 gigawatt-hours (in 1985).[2] By comparison, investor-owned utility generation was 1,487,000 gigawatt-hours in 1975 and 2,213,000 gigawatt-hours in 1991.[3]

The volume of wholesale power transactions among utilities increased during this period in part because of the emergence of excess generating capacity in parts of the United States, regulatory delays in power plant construction in other areas, and de facto market-based pricing for bulk power transactions between investor-owned utilities (Joskow 1993, 26; Niskanen 1992, 12). Until 1985, purchases also increased because rising costs for oil and gas generation made it economical to purchase energy from coal power plants. The Energy Information Administration (1983, xi) concluded that the key factor explaining regional differences in the ratio of bulk power transactions to generation was access to coal-fired and, in the Pacific Northwest, hydroelectric power plants. Nationally, utilities with primarily coal-fired generating capacity tended to be net suppliers of bulk power, while utilities with primarily oil- and gas-fired generating capacity tended to be net purchasers.

Another development has been an increase in wheeling of bulk power by investor-owned utilities. The amount of electric energy wheeled by investor-owned utilities increased from 67,000 gigawatt-hours in 1975 to 228,000 gigawatt-hours in 1991.[4]

NONUTILITY GENERATION

In 1991 investor-owned utilities accounted for 73 percent of electrical generating capacity and 71 percent of electricity generated in the United States. Government entities and cooperatives accounted for 21 percent of capacity and 20 percent of generation.[5] Recently, increasing shares of capacity and generation have been accounted for by private nonutility generators, which have become important competitors in the market to supply long-term bulk power. Nonutility generators include: (1) power plants that meet the technology, fuel, and ownership criteria to become "qualifying facilities" under the Public Utility Regulatory Policy Act of 1978; (2) "independent power producers" and "exempt wholesale generators" that use non-rate-based capacity to generate power exclusively for the wholesale market but that are not qualifying facilities; and (3) other industrial plants that produce electric power in whole or in part for their own use.

Qualifying facilities are exempt from wholesale rate regulation under the Federal Power Act and from regulation under the Public Utility Holding Company Act, while exempt wholesale generators are a new category given exemption from the latter act by the Energy Policy Act of 1992. At the end of 1989, there were only three independent power producers; all were generating units that had been spun off by investor-owned utilities that were not permitted to include them in their rate bases (Joskow 1991, 67). Investor-owned utilities sometimes have ownership interests in qualifying facilities (up to 50 percent), independent power producers, and exempt wholesale generators.

In 1991 nonutility generators accounted for 6 percent of U.S. generating capacity and 9 percent of generation, both up from 3 percent in 1979. Nonutility generators accounted for 20 to 25 percent of the net increase in capacity and generation between 1979 and 1991. Nonutility generators installed more net generating capacity during 1990 and 1991 than did all electric utilities. Half of nonutility generating capacity is in the chemical, paper, and petroleum refining industries (Energy Information Administration 1993, 5).[6]

The growth of nonutility generators was promoted by developments that overloaded the traditional regulatory process. For most of the twentieth century, the real costs and prices of electricity fell because of reductions in real oil prices, improvements in technology, and increasing exploitation of economies of scale. That era ended two decades ago. From the early 1970s to the early 1980s, rate regulation prevented utilities from raising their prices as fast as their costs, especially for fuel, interest, and emissions control. Beginning in 1985, regulators excluded from rate bases substantial amounts of utility investment, including cost overruns associated with nuclear power plants and capacity additions that turned out to exceed growth in demand. This experience discouraged capacity investments by regulated utilities and thus encouraged the development of capacity by nonutility generators. Nonutility generators that are not subject to traditional cost-based regulation are able to retain profits earned by performing above expectations.

Another explanation for the increasing importance of nonutility generators was a reduction in the cost advantage of large generating facilities. This resulted in part from increased risks for large units with long lead times because of growing uncertainty about demand. Also, there have been improvements in natural gas combined cycle and combustion turbine technologies that operate on a smaller scale.[7]

In addition, the Public Utility Regulatory Policy Act of 1978 encouraged entry by nonutility generators. The act enabled cogenerators and small power producers known collectively as "qualifying facilities" to avoid traditional state and federal regulation. Qualifying facilities include (1) cogenerators, which must supply at least 5 percent of their energy output in the form of useful thermal energy (for example, steam used in production of paper or chemicals) and (2) generators using renewable fuel sources. Ownership of a qualifying facility by electric utilities cannot exceed 50 percent.

Qualifying facilities also benefited from a requirement that investor-owned utilities buy power from them at prices determined by the states to reflect the utilities' "avoided costs" rather than the qualifying facilities' costs of service. States have determined avoided costs in various ways, and early on often set the prices above the purchasing utility's true alternative costs. This led to excessive prices and capacity for qualifying facilities (Office of Technology Assessment 1989, 64). Presumably, these problems have been reduced as a result of increasing reliance on competitive bidding to determine suppliers and avoided costs.

Virtually all of the increase in capacity of nonutility generators since the Public Utility Regulatory Policy Act of 1978 has been by qualifying facilities, predominantly cogenerators. Of nonutility generating capacity at the end of 1991, 73 percent was placed in operation since the act was passed. Ninety-one percent of this recent capacity was added by qualifying facilities. Seventy-four percent of qualifying facility capacity is at cogenerators.

Nonutility generators typically provided power solely for their own use prior to 1980. Since 1980 sales of power by nonutility generators to utilities increased from 7,600 gigawatt-hours (11 percent of nonutility generation) to 136,600 gigawatt-hours (50 percent of nonutility generation) in 1991. In 1991 nonutility generators supplied 5 percent of the electricity available for distribution by the electric utility industry.[8]

The experience of qualifying facilities that have won bids indicates that non-rate-based generators can supply new capacity that is competitive with the rate-based capacity of investor-owned utilities (U.S. Department of Energy 1991/1992, 25).[9] However, qualifying facility status is limited to plants with certain technologies and fuels. In many situations, efficient generating plants would not meet the criteria to be qualifying facilities. For example, cogeneration is not efficient in an area with limited potential for industrial use of thermal power. One consequence is that even where a stand-alone power plant using traditional fuel would be efficient, there has been an incentive to waste resources by using cogeneration technology and setting up ancillary facilities (greenhouses, for example) to use thermal energy so that the power plant will be treated as a qualifying facility.

PUHCA REFORM

Until 1992, power plants that were not eligible for qualifying facility status faced entry barriers imposed by the Public Utility Holding Company Act of 1935, and they had limited access rights to the transmission services required to reach potential purchasers. The Energy Policy Act of 1992 removed the restrictions imposed by the Public Utility Holding Company Act. The 1992 act also authorized the Federal Energy Regulatory Commission to order provision of transmission service.

The original rationale for the Public Utility Holding Company Act was to prevent utilities from employing complex holding company structures to evade regulatory scrutiny of their rates and financial condition. One way the 1935 act attempted to do this was by reinforcing the pattern of utilities operating as contiguous, integrated systems that owned and operated the generating facilities that served their native loads (that is, their retail and requirements customers).

The Public Utility Holding Company Act generally had the effect of limiting competition to build new capacity to supply bulk power markets to (1) cost-of-service rate regulated capacity owned by the local electric utility, (2) publicly owned entities, and (3) qualifying facilities (which are exempt from the act).[10] Other potential suppliers—nonutilities and nonlocal investor-owned utilities—that would have had lower costs were excluded by the act or by the threat that ownership of a non-rate-based power plant selling electricity at wholesale would subject all the activities of the owner to burdensome regulation under the act.

A report by the U.S. Department of Energy (1991/1992, 25) describes two interesting cases that illustrate the costs of the act's entry restrictions. In one, "a potential 240-megawatt IPP [independent power producer] was selected by Virginia Power in its 1988 solicitation, but the project was ultimately abandoned because the developer was unable to obtain an exemption from the Public Utility Holding Company Act.

Virginia Power replaced this project with a QF [qualifying facility] of comparable size, but its long-term average annual cost was $35 per kilowatt-year higher, for an annual efficiency loss of $8.4 million." Based on these two cases, the report assumes that the benefits from Public Utility Holding Company Act reform would amount to $30 per kilowatt-year (0.34 cents per kilowatt-hour) for new base-load and intermediate capacity.

The Energy Policy Act of 1992 opened the way for "exempt wholesale generators" to compete in supplying wholesale power to utilities. It did this by exempting facilities engaged exclusively in wholesale power supply from the requirements of the Public Utility Holding Company Act. The rates of these "exempt" wholesale generators remain subject to regulation by the Federal Energy Regulatory Commission.

The 1992 act allows investor-owned utilities, including holding companies, to own exempt wholesale generators. The new law further allows electric utilities to purchase energy from affiliated exempt wholesale generators, but only under certain conditions. Every state commission with jurisdiction over the utility's retail rates must determine (1) that it can adequately regulate the transaction, (2) that the affiliated exempt wholesale generator does not have any unfair advantage, and (3) that the transaction will benefit consumers, is in the public interest, and does not violate state law. Some states prohibit utilities from purchasing power from non-rate-based affiliates.

Even with the Energy Policy Act's limitations, however, utilities with regulated monopoly distribution systems are likely to be allowed to expand further into non-rate-based generation. This will facilitate evasion of regulation and foreclosure of competition. We believe investor-owned utilities subject to cost-of-service rate regulation should not be allowed to own exempt wholesale generators. If they are, they should not be allowed to purchase from them. Our reservations about utility ownership would be reduced substantially if effective reform of rate regulation accompanied the introduction of competition, because the problem of evasion of regulation arises from *cost-based* regulation of prices (Braeutigam and Panzar 1989; Owen 1993a, 23–24).

REGULATION OF WHOLESALE POWER PRICES

Prices play a critical role in determining whether resources will be used efficiently and not be wasted. For resources to be allocated efficiently in the generation of electric power, several conditions must be met:

- Any level of power that is produced in a market must be generated at the plants with the lowest variable costs. We will assume that if generating costs are minimized, the supply of power to the market is given by the Marginal Cost curve in Figure 3.1.
- The output of power must be expanded as long as buyers' willingness to pay for the power, which is indicated by the height of the Demand line in Figure 3.1, exceeds the marginal cost of supplying the power, which is indicated by the height of the Marginal Cost curve. This implies that the output of power must be Q_e.
- The power that is produced must be used by the buyers that are willing to pay the most for it. This implies that the demands for power that are satisfied must be those indicated by the Demand line in the output range from zero to Q_e.

Figure 3.1: Marginal Cost Pricing

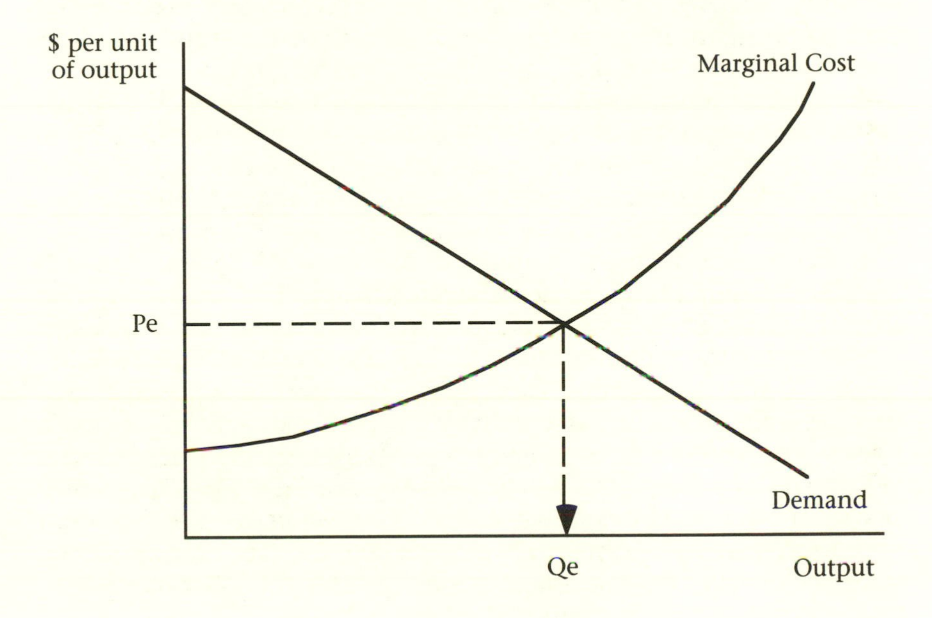

The simplest way to satisfy these conditions for an efficient allocation is to price power at P_e per unit. At that price, the generators with the lowest costs are willing to supply Q_e units of power. All buyers willing to pay the cost of the most expensive unit of power produced would be able to satisfy their demands, while no power would be used by buyers who were unwilling to pay the cost of the most expensive unit of power produced.

Absent market imperfections of the type discussed in Chapter 1, in a competitive market the Marginal Cost curve is also the supply curve, and market-based prices determined by supply and demand will equal P_e. Thus competitive markets achieve marginal-cost pricing, that is, prices are equal to the level of marginal costs at the efficient level of output.

An important feature of marginal cost pricing is that efficient prices will vary with demand. Thus, if pricing is efficient, prices should be lower during off-peak hours of the day and seasons of the year than during peak hours and seasons.

Pricing for bulk power that departs from these marginal cost principles limits efficient power transactions. When prices exceed marginal costs, buyers are deterred from making efficient purchases. This occurs, for example, when sellers exercise market power. If prices are constrained by regulation to be below the level of marginal costs at the efficient level of output, sellers are deterred from making efficient sales. Competition between two suppliers charging prices based on embedded costs rather than marginal costs could lead a purchaser to buy power from the supplier with the higher marginal generating costs.[11]

In practice, when generating capacity is fixed and not fully utilized, the largest component of the marginal cost of energy is the cost of fuel. If generating capacity is fully utilized, the marginal cost of energy for use A is its opportunity cost—the value of that energy in the use that would have to be forgone if the energy were allocated to use A.

Marginal cost pricing principles can also be used to achieve an efficient allocation of firm power—essentially, capacity rights—assuming that associated energy is priced according to marginal cost principles. When generating capacity is fixed, the marginal cost of capacity rights for use B is the value of those capacity rights in the use that would have to be forgone if the capacity rights were allocated to use B. Suppose Utility X is considering the sale of capacity rights for a period of one year to Utility Y. Such a sale might have any of a variety of opportunity costs to Utility X. For example, Utility X might be required to forgo sales to Utility Z of capacity rights for periods shorter or longer than a year, or Utility X might be forced to make greater purchases of capacity to meet its own requirements.

Pricing in Practice

Federal Energy Regulatory Commission regulation of wholesale power prices varies depending on the nature of the parties to the transaction. Transactions between investor-owned utilities and their requirements customers (for example, municipal utilities located within their control areas) and transactions among affiliated entities that might facilitate evasion of retail price regulation are regulated based on traditional cost-of-service principles. In the case of requirements transactions, presumably one objection to marginal cost pricing would be that the utility would not be able to pass along to requirements customers the costs of past investment mistakes.

Wholesale power transactions between investor-owned utilities have been facilitated by loose Federal Energy Regulatory Commission regulation of prices for these transactions. There has been extensive market-based pricing for not only economy energy but short- and medium-term capacity and energy transactions between investor-owned utilities for many years (Joskow 1989b, 189–90; 1993, 26). According to Joskow: "Bulk power prices associated with what the FERC Staff typically refers to as 'coordination service' are *de facto* determined primarily by competitive market forces rather than by rigid cost-of-service principles."[12]

Although utilities are generally required to provide support based on costs for prices for firm power, the Commission has allowed a variety of cost-based formulas. It is likely that competition plays a role in determining which formula is used (Acton and Besen 1985, 22–23). This is another reason that competition plays a role in determining prices.

For example, capacity charges for long-term firm power have been based on system average costs or on the costs of specific units, and capacity charges for short-term firm transactions have been loosely based on the average fixed costs of the units expected to provide the service. Capacity charges have been collected as a fixed fee or as a variable charge based on actual purchases of energy. In the case of unit power transactions, the Commission has allowed utilities to use the cost of debt issued during the construction of the plant in question or the system average cost of debt. Energy charges for firm power may be based on system average costs or incremental costs of specific units, and may or may not include an adder or allowance to cover "nonquantifiable costs" (Earley 1984, 23–33, 65).

The Commission has also approved prices for wholesale power transactions among investor-owned utilities that are explicitly based on factors other than the seller's costs. It has approved uniform rates, rather than rates based on the costs of each selling utility, for short-term and limited-term transactions involving different utilities. In the case of economy energy transactions between investor-owned utilities, the Commission allows prices based on the gains from the transactions. For example, the Commission allows the selling and purchasing utilities to share savings (the difference between their marginal costs of generation, after transmission losses and any wheeling charges) on a 50–50 basis.

The scope for competition to determine rates is further increased because the Commission treats cost-based prices as price ceilings and allows lower prices. According to Tenenbaum and Henderson (1991, n. 1):

> Typically, the Commission has allowed sellers to charge a price up to the sum of actual or projected operating costs plus an adder equal to the embedded capital cost of one or more units on their system for off-system sales. The actual (i.e., out of pocket) cost of generating electricity from an existing generating unit is mostly incremental fuel cost which is usually significantly lower than this allowed ceiling. To maximize their pricing flexibility, some sellers have filed rate schedules that give them the right to charge any price between out of pocket costs and the allowed price ceiling. This gives them the ability to charge prices that reflect changing market conditions.

Similarly, according to Earley (1984, 64), "The general standard used by the Commission in approving capacity or reservation charges has been that they not be greater than the embedded capacity costs reasonably allocable to the sale or service at issue." Also, sellers have been allowed to set capacity charges below embedded costs for unit power when there is excess capacity (Earley 1984, 65–66). In the case of energy charges, "the amount of the split-savings adder is sometimes used as a point of departure for downward pricing flexibility for economy energy sales. The Commission has approved rate schedules that provide for a price that is 'mutually acceptable' to the parties. Any price less than or equal to the split-savings price can be charged without any additional filings with the Commission" (Earley 1984, 75–76). There is also room for competition to influence nonprice terms of transactions.

The Commission went a step further in the western United States under the Western Systems Power Pool. Beginning in 1987, the Commission approved an experiment in which prices for short-term energy and transmission were deregulated, subject to price caps that were sufficiently high so that they were not a relevant constraint (Office of Technology Assessment 1989, 143; Pace 1990; Federal Energy Regulatory Commission 1990a, 46 Tr. 7207 (Joe D. Pace)). By one estimate, during 1988–90 the experiment led to annual savings of $35 million (Strategic Decisions Group 1991). In 1991, on the grounds that transmission conditions were inadequate to mitigate market power in the absence of open access to transmission, the Commission required that market-based rates be replaced by cost-based rate caps for short-term sales of power and transmission. Prices for power sales are to be based on pool-average costs plus a return, although individual companies can use their own higher costs with the Commission's approval (*Electric Utility Week*, June 3, 1991, 2; July 1, 1991, 9).

In addition, since 1987 the Commission has granted market-based pricing for wholesale power sales and transmission on a case-by-case basis (Tenenbaum and Henderson 1991; Joskow 1992). Applicants must convince the Commission that there are no problems related to self-dealing when transactions involve affiliated firms and that there are no problems relating to exercise of market power. In principle, parties to an affiliate deal might convince the Commission that the terms of a transaction were comparable to those in transactions among non-affiliated companies. However, according to one interpretation, a 1991 Commission order relating to Boston Edison "suggests that the only way to win market rates for an affiliate deal is for the affiliate to win a competitive solicitation conducted by the utility" (*Electric Utility Week*, June 3, 1991, 6). The Commission has given a number of investor-owned utilities blanket authority to sell bulk power at market-based rates when the utilities agreed to make wheeling service available under "open access" transmission tariffs (Federal Energy Regulatory Commission 1992b).

Notwithstanding the potential benefits of both open access to transmission and market-based pricing, the Commission's decisions on applications for market-based pricing of bulk power have been based on a flawed methodology for analyzing market power. The Commission appears to have rejected market-based pricing in some situations where market power was unlikely to be exercised. Also, its methodology for evaluating market power could lead it to conclude that the exercise of market power was unlikely in situations where problems actually did exist (Harris and Frankena 1992).

TRANSMISSION ACCESS AND PRICING

Increasing interest in competition in generation has focused attention on the transmission facilities that link sellers and buyers of bulk power. Issues of access to transmission services, transmission pricing, investment in new transmission facilities, and competition among suppliers of transmission services now play a central role in regulatory matters.

Transmission issues have been the focus of investigations into the competitive effects of several recent mergers between investor-owned utilities. Transmission conditions are important determinants of whether suppliers of wholesale power or owners of transmission facilities can exercise monopoly power over purchasers of bulk power, as well as whether buyers of wholesale power or owners of transmission facilities can exercise monopsony power over sellers of bulk power.

In addition, transmission access and pricing are important determinants of the costs of power. In the short run, they influence whether existing generation and transmission resources are used efficiently to produce the services for which potential users are willing to pay the most. In particular, they influence whether power is supplied from the lowest-cost generating facilities. In the longer run, the pricing and availability of transmission influence whether generating capacity is minimized and whether new generating facilities are located appropriately and take advantage of economies of scale. Access conditions and prices also affect incentives to expand the transmission system.

Pricing Principles

To achieve efficient use of an existing transmission system, prices for both rights to use transmission capacity and the actual transmission of energy should be equal to the opportunity costs.

Opportunity costs arise when provision of transmission service makes it necessary for the host utility, or a customer to which the host would otherwise supply power, to rely on higher cost sources of power. For example, provision of transmission service might force the host utility to reduce its reliance on its most efficient generating plants to serve its native load. Alternatively, provision of transmission service might require the host utility to forgo purchases of low cost power or to forgo off-system sales. Again, the result might be that higher cost generation would replace lower cost generation.

One problem with charging opportunity cost prices for transmission service is that the revenues generated by such prices may be insufficient to cover the replacement costs of the transmission facilities, and hence may provide inadequate incentives to invest in transmission capacity. (In computing revenues here, it is assumed that the owners of transmission capacity also pay themselves prices equal to the opportunity costs for using the system.) For example, transmission systems are subject to economies of scale. If the capacity of a system that is subject to economies of scale is at the efficient level, prices that are equal to opportunity costs will be below average costs. As a result, revenues will not cover the total costs of the system (Frankena 1979, chap. 5; 1982, 91–95).

Regulators may find that the way to cover such a revenue gap with the least distortion in resource allocation would be to require that potential purchasers of wheeling service pay a subscription fee for the right to wheel energy at opportunity-cost prices. The alternative may be to permit transmission charges that exceed opportunity costs.

The Commission's Pricing Policies

The Federal Energy Regulatory Commission has generally based prices for both firm and nonfirm transmission service on the average rolled-in embedded cost of the transmission system of the host utility (Holmes 1983; Federal Energy Regulatory Commission 1993b). Charges for transmission losses have been based on average rather than marginal losses (Federal Energy Regulatory Commission 1989a, 61). The Commission has applied "postage stamp" rates rather than basing rates on specifics of the service in question, such as the distance that power is transmitted or the locations of generating plants and loads involved in a particular transaction.

Notwithstanding regulation, competition plays an important role in constraining transmission prices. The Commission has allowed flexibility in pricing of transmission service between investor-owned utilities. For economy energy transactions, a wheeling utility can charge 15 percent—and, on occasion, up to one-third—of the gains from trade, that is, the difference between the incremental generating costs of the selling utility and the buying utility. A transmitting utility can increase its share of the gains from trade to 50 percent by purchasing power from the selling utility and simultaneously selling power to the buying utility, because each transaction is subject to a 50–50 sharing of the gains between the buyer and seller (Federal Energy Regulatory Commission 1989b, 61–62, 120).

In fact, if competitive conditions permit, a transmitting utility would have the flexibility to increase its share of the gains from trade to substantially over 50 percent, because the initial seller of the power is free to negotiate a share below 50 percent of the gains from the initial transaction. Hughes testified that Utah Power and Light was able to avoid the Commission's regulation of transmission rates and to exercise market power in transmission by engaging in the purchase, transmission, and resale of bulk power (Hughes 1988, 7).

There are other reasons that transmission prices are influenced by competition. In the case of wheeling, rates may or may not be subject to adders to cover "nonquantifiable costs," and there have been cases where the wheeling utility was allowed to use safe-harbor rates set by the Commission. Wheeling utilities also have discretion in whether to file for increases in regulated rates.

Recent Developments in Pricing

The Commission has now become more open to opportunity-cost pricing for transmission service. The Commission's revised order in the Northeast Utilities/Public Service of New Hampshire merger case opened the way for transmission rates that include "legitimate, verifiable opportunity costs" (Federal Energy Regulatory Commission 1992a, 61,203). Also, when capacity is expanded to accommodate wheeling, the Commission will accept pricing based on the incremental costs of expanded capacity rather than systemwide average costs.

In a case involving Pennsylvania Electric, the Commission decided that the utility could charge whichever was higher, its embedded-cost rate, or its opportunity costs up to the incremental cost of expanding its system (*Electric Utility Week*, Mar. 23, 1992, 11–12). Utilities providing transmission service argued that if they could charge only the higher of these two costs, the charge should be computed separately for each hour. The Commission ruled, however, that the comparison should be over the life of the contract, which almost inevitably leads to a lower average rate. One commissioner argued that the result would be embedded-cost pricing (*Electric Utility Week*, Oct. 19, 1992, 12).

The Commission's approach to transmission pricing is indicated in a ruling involving Louisville Gas and Electric. The Commission accepted an open access transmission tariff that provides for Louisville Gas and Electric's wheeling rate to be the highest of the following: its embedded cost rate, with a cap of $0.95 per kilowatt-month; its opportunity costs, capped by the cost of expanding its system; or the costs of expanding its system if it makes investments for this purpose (*Electric Utility Week*, Jan. 25, 1993, 2).

Expansion of Capacity

A closely related issue is the obligation of investor-owned utilities to expand their transmission capacity to accommodate wheeling. Investor-owned utilities with market power have an incentive to exercise their market power not only by charging high prices for, or denying, wheeling service but also by limiting the capacity of their transmission systems. Opportunity cost pricing gives a utility with market power an incentive to limit the capacity of its transmission system in order to raise its transmission rates, revenues, and profits. The Federal Energy Regulatory Commission has responded by capping opportunity-cost rates at the cost of expanding the system, and in at least one case by giving third parties priority over the host utility. Of course, any pricing system that does not permit a utility to earn a competitive, risk-adjusted rate of return on investments in capacity expansion will also limit investments.

An important issue is how to structure investor-owned utilities' incentives and create obligations so that these utilities will undertake efficient expansions of their transmission systems.

Where market power in transmission is sufficient to justify the costs of regulation, it makes sense to require owners of transmission systems to expand capacity if beneficiaries will pay the appropriate long-run costs. Absent uncertainty about demand, Alger and Braman (1993) find that this would entail, per unit of capacity, the higher of (1) the average cost of total capacity at the expanded capacity level and (2) the average cost of the additional capacity. Alger and Braman find the price would be higher when there is uncertainty about demand.

One problem is that public utility commissions that focus on effects on utilities and customers within their states may engage in beggar-thy-neighbor practices. They may allow investor-owned utilities to limit their transmission capacity and exercise market power over out-of-state entities, provided the profits are shared with local customers.

Native Load Customers and Shareholders

Access conditions and pricing affect the distribution of benefits of the transmission system between native load customers and the shareholders of the host system, on the one hand, and the customers and shareholders of the entities supplying and purchasing power that would be wheeled, on the other. Under pressure from state regulators, among others, in the merger case involving Northeast Utilities and Public Service of New Hampshire, the Federal Energy Regulatory Commission revised its draft order to provide that transmission access conditions should not harm native load customers.

Some believe that firm transactions by other users should have priority over nonfirm transactions by the host system. This viewpoint is embodied in the Commission's orders in two merger cases: Utah Power and Light/Pacific Power and Light and Northeast Utilities/Public Service of New Hampshire mergers. According to a former commissioner: "In *Utah* the rationale for the measure centered on the need to remedy anticompetitive consequences of the merger, in that the Commission arguably needed to create a preference for third party suppliers to ensure that the merged company would construct additional capacity on a timely basis to meet demonstrated transmission demand" (Federal Energy Regulatory Commission 1989b, 3).

No one seems willing to propose transmission access measures that they acknowledge would reduce reliability. This leaves the important question of what reliability standards are to be used, how effects on reliability are to be measured, and how measurements will be verified.

Mandatory Access

Until 1992, unless a utility was applying for approval for market-based pricing or a merger, the Commission's power to order transmission access was very limited. The Federal Power Act prohibited the Commission from ordering access unless it determined that doing so "would reasonably preserve existing competitive relationships."

The Energy Policy Act of 1992 granted the Commission the authority to order wholesale (but not retail) transmission service, including any enlargement of transmission capacity necessary to provide such service. Sixty days after requesting transmission service, any generating facility that cannot reach an agreement with a utility can apply to the Commission for a transmission order. The Commission may not issue such an order if it finds that it would unreasonably impair reliability, and unless it is in the public interest. The Commission is not required to make any finding regarding market power. The act requires that the rates for mandated transmission service be "just and reasonable" and also that transmission service "rates, charges, terms, and conditions shall promote the economically efficient transmission and generation of electricity." Other provisions relating to transmission rates—what one former member of the Federal Energy Regulatory Commission has characterized as "a mishmash of language . . . that, on the surface, seems to give lobbyists for every interest a claim to victory" (Stalon 1993, 34)—appear to give the Commission the authority to continue its present pricing practices or to change to more efficient pricing.

The 1992 law is silent on whether transmission rights purchased pursuant to an order by the Commission can be resold. Clearly, purchasers should be allowed to resell rights. If resale is not permitted, transmission capacity will not be allocated to the most valuable uses.

Some have predicted that the Energy Policy Act will reduce investments in transmission capacity and thus frustrate the growth of competition in bulk power (*Electric Utility Week,* Nov. 23, 1992, 1, 12–13). Utilities might be less interested in making transmission investments, given the Commission's new powers over access and its pricing policies. And state regulators, who must approve new transmission facilities and their inclusion in the retail rate base, could resist transmission investments that would benefit consumers in other states but that might raise retail rates in the host state.

The Energy Policy Act of 1992 opens the way to increase regulating transmission on the assumption that mandatory wheeling will increase competition among generators, which would permit a reduction in regulation of generation. However, Gilbert et al. (1993) conclude that exercise of the Commission's new powers to mandate wheeling would not be likely to result in more efficient use of existing generating capacity. This conclusion is based on crude evidence that, given the Commission's loose regulation of short- and medium-term transactions, the wholesale market was working relatively well prior to the 1992 act.

Massachusetts and a number of other states have mandatory intrastate wheeling, at least in connection with their competitive bidding programs. Wisconsin's public utility commission has the authority to order a utility to open its transmission system to other utilities. State authority to order wheeling, however, may be preempted by the Federal Energy Regulatory Commission's authority (Office of Technology Assessment 1989, 57, 65, 221).

As laws and regulations are changed to increase third-party access to transmission, it is becoming increasingly important to have efficient pricing for transmission services. If transmission access is limited, transmission prices will not have a substantial effect on decisions regarding the location of new generating plants and investments in additional transmission facilities. As transmission access is increased, inefficiencies in transmission pricing can be expected to cause more important distortions in resource allocation. The Commission has recently expressed a willingness to reconsider its pricing policies, including ways to take account of parallel path flow (Federal Energy Regulatory Commission 1993b, 1993c).

Proposals for changes in transmission access and pricing that are more fundamental than those considered here would often require changes in the way high-voltage transmission networks are owned (Smith 1991; Hogan 1992).

Analogy to Telecommunications

The analogy to access issues involved in telecommunications is instructive. Access policy played a significant role in telephone competition from 1970 until the breakup of the old Bell System in 1984. The Federal Communications Commission, promoting limited forms of long-distance competition, required local telephone distribution companies to provide "equal access" to the firms, such as MCI, that sought to compete with AT&T in supplying long-distance service. At least partly because they were owned by AT&T, the local firms resisted these orders and for several years succeeded in delaying and impeding access despite a series of court orders upholding the Federal Communications Commission's mandate. Numerous technical arguments regarding the impossibility of equal access and the problem of "providers of last resort" were mooted. Many activities of local distribution companies were identified by the Justice Department as anticompetitive, providing part of the basis for the settlement that broke up the Bell System. Since 1984, with local telephone distribution companies no longer part of the Bell System, disputes involving access have subsided, and competition in long-distance service appears to be robust.

CONCLUSION

Antitrust analysis of electric utility mergers is important because consumers benefit from competition, and competition plays a significant role in determining prices and resource allocation for electric power, particularly at the wholesale level. Furthermore, the role of competition can be expected to increase because of legislative and regulatory changes.

Among the important developments during recent decades that indicate the importance of competition is the expansion of wholesale power markets. Bulk power purchases by investor-owned utilities from other utilities and from nonutility generators have increased, as has the volume of wheeling transactions. Furthermore, because regulation of prices for short- and medium-term bulk power transactions among utilities is loose, competition can determine prices. Long-term power transactions, including utility purchases from nonutility generators, are often subject to competitive bidding.

In 1992 new federal legislation opened the way for substantial changes in the role of competition in the electric power industry. The Energy Policy Act substantially reduced restrictions on entry into the generation and supply of long-term bulk power by nonutility generators. The act also gave the Federal Energy Regulatory Commission new powers to order utilities to provide transmission service to other entities that generate power.

Key issues in coming years will relate to how the Federal Energy Regulatory Commission uses its new powers, including whether it bases new rules and decisions on sound analysis of competition issues. An emerging issue is whether the role of competition in retail distribution will be increased by changes in laws or regulations that provide for retail wheeling to large purchasers of electricity.

NOTES

1. A gigawatt is a million kilowatts. The 1975 figure is for all investor-owned utility purchases (Energy Information Administration 1985, table 28) minus electric utility purchases from industrials (Edison Electric Institute, *Statistical yearbook of the electric utility industry 1988*, table 9). The 1991 figure is for investor-owned utility purchases from utilities (Energy Information Administration 1992, table 29).

2. Energy Information Administration 1992, table 29, and Energy Information Administration, *Financial statistics of selected investor-owned electric utilities*, annual, for earlier years. According to the Energy Information Administration: "Interchanges are generally short-term transactions that occur in response to temporary load and cost conditions, while...purchases tend to be transactions based on long-run cost differences." Interchanges were only 60,000 gigawatt-hours as of 1965, when generation was 807,000 gigawatt-hours. The ratio of interchanges to generation began to increase in the late 1960s when oil prices began to increase. Energy Information Administration 1983, 1, 3.

3. Edison Electric Institute, *Statistical yearbook of the electric utility industry 1988*, table 8, and *1991*, table 15.

4. Energy Information Administration 1983, 1992, table 29, and *Financial statistics of selected investor-owned electric utilities*, annual, table 29, for earlier years; Edison Electric Institute, *Statistical yearbook of the electric utility industry 1991*, table 8. The increase in wheeling was interrupted by a drop from 197,000 gigawatt-hours in 1985 to 133,000 gigawatt-hours in 1986; at least in part this may reflect a change in reporting. Federal Energy Regulatory Commission 1989a, 191–94.

5. Edison Electric Institute, *1991 Capacity and generation of non-utility sources of energy*, tables 1 and 15.

6. Joskow 1989, 166; Edison Electric Institute, *1989, 1990*, and *1991 Capacity and generation of non-utility sources of energy*, table 1, and *1986 Capacity and generation of non-utility sources of energy*, table 20; Energy Information Administration, *Annual outlook for U.S. electric power 1991*, xi, 19, and 1993, vii, 5.

7. Energy Information Administration, *Annual outlook for U.S. electric power 1990*, 10, and Congressional Research Service 1991, 12, 19–21.

8. Edison Electric Institute, *1991 Capacity and generation of non-utility sources of energy*, tables 2 and 15, and *Statistical yearbook of the electric utility industry 1986* and *1991*, table 9.

9. This interpretation of the evidence is not without its critics. See, for example, Swidler 1991, 14–18, 40.

10. For a detailed explanation of the Public Utility Holding Company Act, see U.S. Department of Energy 1991/1992.

11. Joskow (1991, 84–85) argues that, in addition, state regulation may need to be reformed to provide utilities with incentives to minimize costs in choosing between utility-owned generation and purchased power.

12. Edison Response to Federal Energy Regulatory Commission Staff Data Request (Case-in-Chief) OEP-2-81.b.1. Federal Energy Regulatory Commission Docket No. EC89-5-000. Southern California Edison and San Diego Gas and Electric merger.

4

Market Definition in the Electric Power Industry

Will a merger significantly increase the likelihood that market power will be exercised? This is the principal issue addressed by an antitrust analysis. To answer the question one must define the markets in which monopoly or monopsony power might be exercised. This chapter explains the principles used to define relevant antitrust product and geographic markets. The following chapter explains how to answer the questions that come next in an antitrust analysis: What would be the effect of the merger on the concentration of suppliers in the market? To what extent would new entry, or the threat of entry, constrain the exercise of market power?

Market definition plays a central role in antitrust analysis. Decisions regarding what is included or not included in a market can affect whether merging firms are treated as competitors, the market shares of the merging parties, and the level of concentration. Thus market definition is often highly contentious, and the misuse and abuse of market definitions are common in litigation.

To define relevant markets, the first step is to identify the products and geographic areas for which the merging firms compete (that is, horizontal overlaps). The initial identification of overlaps may be tentative because the existence of overlaps will depend in some cases on how broadly or narrowly the market is defined. For example, does transmission from A to C compete with transmission from B to C? Similarly, the vertical relationships between the merging firms should be examined for situations that could give rise to competitive problems (for example, the vertical merger of a regulated monopoly and one of its unregulated suppliers).

Next, it is useful to specify the anticompetitive scenarios that might result from the merger. For example, if a proposed merger takes place, the price of Product X will be increased to customers located in Area Y as a result of the exercise of market power. One reason for specifying the various anticompetitive scenarios is that the relevant markets may depend on what anticompetitive behavior is being considered. For example, if price discrimination among types of customers or geographic areas is possible, then relevant markets are likely to be drawn more narrowly.

Defining markets helps identify the firms that, through coordinated action, might exercise market power. Problems defining markets can be reduced if one focuses on market definition, not as an end in itself, but rather as one step in the process of determining whether some group of customers is likely to be injured through the exercise of market power as a result of the proposed merger. The evaluation should consider the following three issues in sequence: potential victims, demand-size substitutes, and supply-side responses.

1. *Potential Victims*. Based on a hypothesis about how market power would be exercised, identify a specific set of customers that might be victimized by an increase in the price of some product or service (Service X).

2. *Demand-Side Substitutes*. Identify other products or services that are substitutes for Service X from the point of view of these customers. Customers cannot be victimized by an increase in the price of Service X if enough of them can switch to some other product or service (Service Y) without suffering significant adverse effects. If enough customers can switch, suppliers of Service X will not find a price increase profitable.

3. *Supply-Side Responses*. Identify suppliers that can respond to the increase in the price of Service X by beginning to supply either Service X or Service Y. Customers cannot be victimized by current suppliers increasing the price of Service X if enough customers can switch to a new supplier that is not currently producing Service X but that would do so if it could sell its output.

THE MERGER GUIDELINES

The U.S. Department of Justice and the Federal Trade Commission analyze the competitive effects of proposed mergers using a framework that is described in their 1992 *Horizontal Merger Guidelines*. These guidelines state that for antitrust purposes:

A market is defined as a product or group of products and a geographic area in which it is produced or sold such that a hypothetical profit-maximizing firm, not subject to price regulation, that was the only present and future producer or seller of those products in that area likely would impose at least a "small but significant and nontransitory" increase in price, assuming the terms of sale of all other products are held constant. A relevant market is a group of products and a geographic area that is no bigger than necessary to satisfy this test. (U.S. Department of Justice and Federal Trade Commission 1992, Section 1.0)

The key to the *Merger Guidelines* approach to market definition is the hypothetical monopolist test. Suppose one firm gained control of all supplies of a product (or group of products) sold in a specified geographic area. Could that hypothetical monopolist increase its profits by raising prices above the levels that would otherwise prevail? Equivalently, one could phrase the test in terms of hypothetical collusion: Suppose all sellers of a product (or group of products) sold in a specified area could perfectly coordinate their pricing and other behavior. Would they increase joint profits by raising prices above the levels that would otherwise prevail? If the answer to these questions is affirmative, then an antitrust market has been identified.

By contrast, a product and a geographic area in which it is sold would not be an antitrust market if a price increase imposed by a hypothetical monopolist (or coordinated group of suppliers) would lead to sufficient substitution on the part of buyers and responses by other suppliers so that profits would decline. According to the *Merger Guidelines*:

In determining whether a hypothetical monopolist would be in a position to exercise market power, it is necessary to evaluate the likely demand responses of consumers to a price increase. A price increase could be made unprofitable by consumers either switching to other products or switching to the same product produced by firms at other locations. The nature and magnitude of these two types of demand responses respectively determine the scope of the product market and the geographic market. (Section 1.0)

The next step is to identify the firms participating in the market. Under the *Merger Guidelines*:

Participants include firms currently producing or selling the market's products in the market's geographic area. In addition, participants may include other firms depending on their likely supply responses to a "small but significant and nontransitory" price increase. A firm is viewed as a participant if, in response to a "small but significant and nontransitory" price increase, it likely would enter rapidly into production or sale of a market product in the market's area, without incurring significant sunk costs of entry and exit. (Section 1.0)

Firms that are not currently producing the product but that could do so easily are referred to in the *Merger Guidelines* as "uncommitted entrants." Such entrants provide an easy alternative for consumers to turn to if current producers raise prices anticompetitively. Thus uncommitted entrants can provide a powerful check on the market power of producers. If the uncommitted entrants could easily produce the relevant product, then they must already be exerting some competitive influence on the market. Accordingly, it is reasonable to include them in the market.

To determine which firms are uncommitted entrants, the antitrust agencies postulate a "small but significant and nontransitory" price increase for the relevant product. A firm is considered an uncommitted entrant if, given the assumed price increase, it would be likely to enter within one year and if, through the sale of output and assets, it could recover all its costs if it exited within one year of initiating entry (Section 1.32). Firms that would incur significant sunk costs to enter are called "committed entrants" by the *Merger Guidelines*. They are considered to be outside the market.

In order to apply the hypothetical monopolist test, the concept of a "small but significant and nontransitory" increase in price must be quantified. According to the *Merger Guidelines* (Section 1.11), "the Agency, in most contexts, will use a price increase of five percent lasting for the foreseeable future," although "the Agency at times may use a price increase that is larger or smaller than five percent." Department of Justice personnel have indicated that a price increase of 5 to 10 percent is often used (Whalley 1989).

The *Merger Guidelines* methods of defining markets have gained widespread acceptance. Federal courts have used the standards in merger cases, and the Federal Energy Regulatory Commission has relied on the methods in its regulatory proceedings.[1]

Relevant markets are defined with reference to the facts in particular antitrust matters. For example, in some circumstances, short-term transmission of bulk power may be a relevant product market. In other cases, the relevant product market may include both short-term transmission and local generation of short-term bulk power in the destination area. Since the key elements in market definition are demand and supply responses to price changes, antitrust markets may vary with demand and supply conditions. For example, demand for power depends on the time of day and on the season. The cost of producing power in certain areas (for example, the Pacific Northwest) depends on the level of precipitation. For these reasons, geographic markets for short-term transmission services can vary with time of day, season, and weather conditions (Owen 1989, 120–30).

PRODUCT MARKET DEFINITION

When defining product markets in connection with electric utilities, one must distinguish between wholesale power transactions and retail electricity transactions. If there is a bulk power generation or transmission overlap between the merging firms, one must determine whether different types of transactions (for example, long-term and short-term transactions) are in the same product markets. When two competing transmission companies attempt to merge, other issues arise. Does it make sense to analyze unbundled transmission (wheeling) separately from transmission that is bundled with bulk power? To what extent would generation proximate to the purchaser constrain the exercise of market power in transmission? In mergers between electric and gas utilities, one must decide whether retail electricity and gas are in the same product market. There will often be more than one relevant market pertaining to a given electric utility merger.

Bulk Power Versus Retail Transactions

From the point of view of customers, suppliers, and regulators, wholesale or bulk power transactions differ significantly from retail electricity transactions. Therefore, bulk power and retail power generally would not be in the same antitrust market. Participants in the bulk power market are utilities, nonutility generators, local distribution companies, and state and federal power agencies that supply or purchase wholesale power at relatively high voltages. The power is usually resold and delivered at lower voltages to retail customers by electric distribution companies.

Purchasers usually have little if any ability to switch between the bulk power market and the retail market. Some large industrial users may be an exception. The ability of utilities to switch sales between the markets is limited by their obligation to serve retail customers—as well as wholesale requirements customers—at regulated rates. Regulations and lack of distribution facilities prevent independent power producers and qualifying facilities from having direct access to retail customers.

Long-, Medium-, and Short-Term Bulk Power

Utilities plan and operate their systems to minimize costs while satisfying their reliability criteria. This involves a distinction between capacity and energy. First, a utility lines up, through ownership or contracts, firm access (including transmission service) to sufficient generating *capacity* so that it can meet its projected loads with a reserve margin to allow for transmission and generating outages. Commitments concerning capacity are often made many years in advance because of the time required to build major generating and transmission facilities. Utilities, however, also enter into medium- and short-term capacity transactions because they inevitably have shortages and excesses of capacity within their long-term capacity plans.

After obtaining capacity, the utility must obtain the *energy*—the electrons—needed to supply the actual loads on its system. Certain generating facilities must operate at some minimum level to provide capacity efficiently, so utilities obtain a portion of their energy needs (their "baseload") from some of the generating facilities that are providing them with capacity. For this reason and also to diversify risks, utilities sometimes buy energy along with capacity under long-term contracts. Frequently, however, utilities buy a portion of their energy requirements in short-term or spot energy markets to save money. In some cases, this is handled through a power pool, which supplies the energy requirements for a group of utilities from the generating capacity available to pool members that has the lowest energy costs. In other cases, it is accomplished by hourly "economy energy" transactions among independent utilities.

Joskow (1989a, 29–30) has conceptualized "as many as three distinct bulk power supply product markets—the market for short-term (primarily) energy transactions, the market for long-term (planned) capacity and associated energy, and the market for medium-term (excess) capacity and associated energy." He explains that the third product consists of "medium-term (e.g., three to five years) capacity transactions designed to transfer capacity from utilities with unplanned surpluses to utilities with unanticipated or deferrable capacity needs." Because utilities must manage substantial uncertainty and satisfy the requirements of regulators and various pooling agreements, one type of bulk power may not be easily substituted for another.

Numerous complications arise in product market definition for bulk power. We will consider two. First, a long-term transaction (in the sense the term is used by Joskow) involves capacity and associated energy beginning several years in the future. An actual long-term contract involving capacity and energy beginning now and continuing for two decades or longer would actually be a bundle of a short-term transaction, a medium-term transaction, and a long-term transaction. To analyze competition for such a contract correctly, one might need to define three separate product markets. The competitive conditions for each portion could be different, and they could be differently affected by a merger. They should be treated as separate markets if differences in competitive factors—such as concentration, entry conditions, and opportunities for self-generation—make it possible that market power could be exercised in one time frame but not in another.

For example, the number of independent transmission paths might be sufficient to eliminate any significant probability of market power in the short term. Some of these paths, however, might not be available for longer term commitments. Members of the Mid-Continent Area Power Pool can purchase capacity from other members for four years under schedules determined by the power pool.[2] This provision might substantially eliminate any potential for exercise of market power for medium-term power. Entry conditions and the existence of excess capacity could make new entry unlikely in the near term, although in the long term it might be realistic.

Second, transactions involving only capacity (without associated energy) for a given term might be a separate product market in some cases. This could occur in situations in which the marginal cost of supplying capacity was relatively low using facilities (such as inefficient excess generating facilities or oil/gas combustion turbines) that have relatively high operating costs to produce energy.

In litigation, it is not unusual to see market definitions being misused or abused, with the effect of changing the apparent shares of the merging parties or the apparent level of concentration in the market. At least where there is agreement that long-, medium-, and short-term bulk power are in separate product markets, one would not expect "market" shares computed on the basis of *all* bulk power supplies regardless of term to be useful in competitive analyses. Yet the same parties that sponsored Joskow's testimony concerning the distinctions between long-, medium-, and short-term transactions presented one of the merging firm's shares of all power used by a purchasing utility, when in fact that merging firm sold only short-term power (Southern California Edison and San Diego Gas and Electric 1990, 57).

Transmission Service

Bulk power transactions take place in both bundled and unbundled forms; that is, transmission service may be supplied in the same transaction as the bulk power or in a separate transaction. Suppose Utility A wants to obtain bulk power from Utility B, with which it is not directly interconnected, and suppose that Utility C has transmission facilities that connect utilities A and B. Exactly the same transaction might be structured in two different ways.

First, Utility A could purchase the bulk power from Utility B and then separately arrange for Utility C to transmit the bulk power from Utility B to Utility A. In this "unbundled" case, Utility C is supplying "wheeling service" (transmission service for bulk power that Utility C does not own).

Second, Utility C could purchase the same bulk power from Utility B, bundle it with the same transmission service from Utility B to Utility A, and sell the bundle as delivered bulk power to Utility A. In this case, Utility C is "brokering" or "trading" bulk power from Utility B to Utility A. This could take place as two separate and apparently unrelated transactions: Utility C's purchase of bulk power from Utility B, and Utility C's sale of delivered bulk power to Utility A.

Whether it is bundled with the bulk power in a trading transaction or sold separately as wheeling service, the same transmission service is being provided by Utility C in these two cases. Obviously, wheeling service by itself is not an antitrust product market. An attempt by a hypothetical monopolist of wheeling service to raise the price charged for wheeling would lead to the restructuring of transactions as trading, and vice versa. In fact, at least in part because of differences in regulation between bundled and unbundled transactions, wheeling is much less common than trading, at least for utilities in Southern California (Owen 1989, 140).

Any product market that includes wheeling transactions would include the transmission component of bundled sales of delivered bulk power. Notwithstanding this fact, litigants have wrongly defined the product market to minimize the role of merging parties in the provision of transmission services. Throughout the Federal Energy Regulatory Commission proceedings to evaluate the proposed merger between Southern California Edison and San Diego Gas and Electric, the merging parties repeatedly identified "transmission services" with "wheeling services," while putting bundled transactions (including the transmission component) in a separate bulk power market. Given the relatively limited role of wheeling in San Diego Gas and Electric's transmission of bulk power, the effect was to portray San Diego Gas and Electric's role in transmission as orders of magnitude smaller than it actually was (Southern California Edison and San Diego Gas and Electric 1990, 62–63). Yet, in response to questions, all of the merging parties' economic experts agreed that transmission service includes not only unbundled wheeling but the transmission component of bundled transactions.[3]

Applicants' efforts to gerrymander market definitions were dismissed by the California Public Utilities Commission in its decision rejecting the merger. The commission found that there was a "short-term firm and interruptible transmission market" that included "wheeling and the transmission component of sales of delivered bulk power" (California Public Utilities Commission 1991b, 41, 37).

Local Generation

In antitrust cases involving transmission of bulk power, should generation that is located in the purchasing area be included in the relevant market with transmission? This is an empirical question, and the answer will vary from case to case. The answer does not depend solely on the fact that locally generated electricity has the same technical characteristics as imported electricity. It also depends on the *cost* of producing additional power locally compared with the price of such power delivered from outside the area. The key is whether the incremental cost at which additional locally generated power would be available would be sufficiently low. If a monopolist controlled all relevant transmission to the destination area, would the monopolist find it unprofitable to raise the price of transmission service by a "small but significant and nontransitory" amount because buyers would switch to local generation? If so, local generation is in the market.

Purchasing utilities' incremental costs of generation in northern California usually substantially exceed the cost of economy energy from the Southwest (Arizona and New Mexico). The availability of additional energy generated locally would not prevent a transmission monopolist with control over all short-term transmission from the Southwest from exercising market power over transmission from the Southwest to northern California. Therefore, customers in northern California could not escape a transmission price increase by substituting local energy for imported energy. Local generation in northern California was not in the product market relevant to overlaps in short-term transmission for the merger between Southern California Edison and San Diego Gas and Electric.[4] There was no record evidence to support the applicants' assertion that it was. Joskow admitted that the facts may not have supported the applicants' assertion (Federal Energy Regulatory Commission 1990a, 49 Tr. 7678, 7686).

By contrast, in two recent cases involving applications for market-based prices for long-term power contracts, there was some evidence that the option of local self-generation by the purchasing entity would have constrained the exercise of market power by a hypothetical transmission monopolist. There was no apparent reason in these cases to believe that the cost of local self-generation would have been significantly higher than the cost of producing and delivering power from elsewhere. Nonetheless, the Federal Energy Regulatory Commission denied the applications. It claimed the applicants had not demonstrated the absence of market power (*TECO Power Services and Tampa Electric Company*, 52 FERC ¶61,191 (1990) and *Terra Comfort Corporation and Iowa Southern Utilities Company*, 52 FERC ¶61,241 (1990)).

GEOGRAPHIC MARKET DEFINITION

At least three geographic areas or dimensions must be considered in the context of electric utility mergers: (1) the area in which market power might be exercised (that is, the area in which a hypothetical monopolist might impose a price increase on customers); (2) the locations and marketing areas of the merging firms, and of other firms that might be involved in exercising market power; and (3) any inherent geographic dimension of the product itself.

The *Merger Guidelines* define a geographic market in terms of the area in which a hypothetical price increase would be imposed on customers. When suppliers can practice price discrimination, the geographic area in which customers may be subjected to a price increase, and therefore the geographic market, may be smaller than would otherwise be the case (Section 1.22). The *Merger Guidelines* also point out that "a single firm may operate in a number of different geographic markets" (Section 1.2). An example is a retail chain with outlets in many local markets. The Federal Trade Commission has defined relevant local markets in cases involving mergers between chains of grocery stores as well as funeral homes (*The Vons Companies et al.*, 111 FTC 64 (1988), *American Stores Co. et al.*, 111 FTC 80 (1988), and *Service Corp. International,* 5 Trade Reg. Rep. (CCH) ¶23,110 (1992)). Because of price discrimination and because a single firm can operate in more than one geographic market, there is not necessarily a close relationship between geographic markets and the total area in which merging firms operate.

Certain products have an inherent geographic dimension. This dimension may help to define the geographic market if there are limits on substitutability for buyers and sellers. For example, airline trips have geographic origins and destinations. The Department of Justice analyzes airline mergers in terms of city-pairs for airline service (Department of Transportation Docket 43754, Northwest/Republic acquisition, and

Docket 43837, TWA-Ozark acquisition). Electrical transmission facilities also connect certain origins and destinations.

To define a geographic market, one should not always begin with the geographic areas served by the merging firms and expand outward, as applicants and Commission staff have been known to do (Southern California Edison and San Diego Gas and Electric 1990, 40; Federal Energy Regulatory Commission 1990d, 55). In the case of bulk power transmission, the place to start is with an individual buyer. One then asks what alternatives the buyer has.

Geographic Markets for Transmission Services

In the absence of regulation, a hypothetical monopolist that controlled all bulk power generation and transmission in a region could practice price discrimination in supplying bulk power to individual wholesale buyers (or groups of contiguous wholesale buyers) located within that region. Because of the hypothetical monopolist's control over transmission, price discrimination could not be avoided by resales of bulk power among noncontiguous customers. As a result, individual wholesale buyers (or contiguous groups) would be in separate geographic markets. Of course, regulation could constrain the hypothetical monopolist's ability to practice price discrimination. For example, the monopolist might be required to supply power to different wholesale customers, such as full-requirements customers, at uniform prices. In that case, some groups of noncontiguous customers might belong to the same geographic market.

The reasoning would be the same for a hypothetical monopolist that controlled all bulk power transmission if local generation of bulk power in the purchasing area was not in the relevant product market. For the merger of Southern California Edison and San Diego Gas and Electric, local generation in northern California was not in the market with short-term transmission to northern California. To assess the significance of the competition between Southern California Edison and San Diego Gas and Electric in short-term transmission from the Southwest to northern California, it would be appropriate to define the relevant geographic market in terms of short-term transmission to northern California. The remaining question is whether transmission origins other than the Southwest would be in the same market. For example, would the availability of transmission of bulk power from the Pacific Northwest or Mexico to northern California prevent a "small but significant and nontransitory" increase in price by a hypothetical monopolist controlling transmission from the Southwest to northern California?

Both Southern California Edison and San Diego Gas and Electric had long-term contracts to purchase bulk power in Mexico during 1986–96. At the time the merger was being evaluated, however, Mexico apparently did not have additional power to sell. In fact, during summers it was repurchasing the power it had sold to Southern California Edison. In any case, since San Diego Gas and Electric and Southern California Edison were the only entities with the ability to transmit power from the proximate part of Mexico to northern California (without routing it through the Southwest), the addition of transmission from Mexico to northern California to the market would not mitigate any competitive concern arising out of the overlap in transmission between the Southwest and northern California.

Even if transmission from the Pacific Northwest to northern California would constrain the ability of a hypothetical monopolist of transmission from the Southwest to northern California under certain circumstances, there are likely to be periods when purchasers in northern California could not turn to additional transmission from the Pacific Northwest to avoid a "small but significant and nontransitory" increase in price for transmission from the Southwest. First, the ability of the Pacific Northwest to supply competitive power hinges primarily on the level of precipitation. In years with normal or above normal precipitation, during periods such as the spring runoff, relatively cheap power is available in the Northwest (although this is less true under the current policies of the Bonneville Power Administration than it was previously). During such periods, transmission from the Pacific Northwest to California is likely to be used to capacity, and thus there would be no additional capacity to constrain price increases on transmission from the Southwest. Second, during years with below normal precipitation, competitive power is not available in the Pacific Northwest. In fact, in the Pacific Northwest, drought conditions have prevailed in a majority of the years of the past decade. Accordingly, transmission from the Southwest to northern California would be an antitrust market.

The Department of Justice found that short-term transmission of bulk power from the Southwest to northern California was a relevant antitrust market in the Southern California Edison/San Diego Gas and Electric merger case. It also found that the proposed merger would significantly reduce competition in that market (U.S. Department of Justice 1990, 7–11). Moreover, the merger would have adversely affected transactions between Pacific Gas and Electric, located in northern California, and energy suppliers in the Southwest "to the economic detriment of PG&E's ratepayers" (California Public Utilities Commission 1991a, 749, 751).

After using the hypothetical monopolist test to determine relevant antitrust markets, one can identify the firms in the market. In other words, which firms can supply services in the relevant market without requiring the approval of another firm in the relevant market?

Consider the case of short-term transmission from the Southwest to northern California. Which firms, through ownership or contracts, have rights to firm transmission capacity from the Southwest to northern California and could use those rights to provide short-term bundled or unbundled transmission service to others? In the California utility merger case, applicants' economic experts testified that San Diego Gas and Electric was not a significant actual or potential supplier of short-term transmission service from the Southwest to northern California (or anywhere else that Southern California Edison was a supplier). That testimony was flatly rejected as inconsistent with the facts by the California Public Utilities Commission: "SDG&E is an active participant in the market supplying interruptible and short-term firm transmission and its competitive role in that market vis-à-vis Edison is likely to expand in the future, absent the merger" (California Public Utilities Commission 1991b, 37–38).

PITFALLS IN MARKET DEFINITION

In analyzing the overlap between Southern California Edison and San Diego Gas and Electric in short-term transmission from the Southwest to northern California, the merging parties and the Federal Energy Regulatory Commission trial staff made several analytical errors. Each of these errors involved use of an excessively broad and unconcentrated market.

Past Purchases as Evidence of Alternatives

One mistake was to argue that all the firms from which a particular buyer had purchased bulk power in the recent past were in the relevant market. These firms were then considered to be alternatives to Southern California Edison and San Diego Gas and Electric as sources of transmission service (including transmission service bundled with bulk power).[5]

Suppose that Utility A (in northern California) once purchased bulk power from Utility B (in the Southwest). That does not demonstrate that either Utility A or Utility B has (or ever had) rights to transmission from the Southwest to northern California. Previous purchases may have been wheeled by another entity with transmission rights, Utility C (or by Southern California Edison or San Diego Gas and Electric).

Imagine this anticompetitive scenario. Following the merger, Southern California Edison/San Diego Gas and Electric, in collusion with Utility C, raised the price of transmission service from the Southwest to northern California. If it required transmission services available only from Southern California Edison/San Diego Gas and Electric or Utility C, Utility A would not be able to avoid such a price increase by purchasing bulk power from Utility B. In the past, Utility C provided transmission, but it would not provide transmission at competitive prices after the merger if it became a participant in a collusive arrangement to raise prices.

The Western United States as a Geographic Market

During the California utility merger investigation, the Federal Energy Regulatory Commission staff and applicants argued that the Western Systems Coordinating Council area (the western third of the United States plus portions of western Canada and northern Mexico) was the relevant geographic market for *delivered* bulk power, that is, power bundled with transmission (Southern California Edison and San Diego Gas and Electric 1990, 53–54; Baughcum 1989, 22). In a market so expansive, given the large number of utilities, concentration was low. Hence, use of that market would largely have eliminated competitive concerns relating to horizontal overlaps in bulk power.

The entire area of the Western Systems Coordinating Council was defined as a geographic market for the following reasons: the merging firms traded in bulk power throughout the region; there were substantial shipments of bulk power among firms throughout the region; and prices of bulk power in different parts of the region were highly correlated (Southern California Edison and San Diego Gas and Electric 1990, 12, 55; Federal Energy Regulatory Commission 1990d, 33, 42, 48, 58–59).

Even if the preceding points were correct, the definition of the relevant geographic market was not. It was important to investigate the effect of the merger on concentration in control over available transmission capacity in the corridor between the Southwest and northern California. Because that concentration was high, buyers in northern California faced a competitive problem. It was no help to them that concentration in control over total generation, transmission, and/or bulk power sales is low in the whole area of the Western Systems Coordinating Council. As the California administrative law judges recognized, "applicants' approach contemplates no transmission barriers. . . In contrast, other parties note that transmission is crucial, since without it bulk power cannot be delivered" (California Public Utilities Commission 1991a, 777).

Smallest Relevant Markets

Of course, the entire Western Systems Coordinating Council would pass the hypothetical monopolist test used in defining antitrust markets. A hypothetical monopolist of generation, transmission, or bulk power sales in the entire Western Systems Coordinating Council would obviously have market power. Market power, however, also could be exercised in a smaller area, such as the transmission corridor from the Southwest to northern California. This area *also* could pass the hypothetical monopolist test. When evaluating the likelihood that market power will be exercised, it is appropriate to use the smallest geographic area that contains products or services of both merging parties and that satisfies the hypothetical monopolist test (U.S. Department of Justice and Federal Trade Commission 1992, Section 1.21). Quite late in the California utility merger investigation, the Federal Energy Regulatory Commission staff acknowledged that it had failed to analyze the smallest relevant markets (Federal Energy Regulatory Commission 1990c, 14–15).

Antitrust Markets Versus Economic Markets

It is important to use appropriate criteria in defining relevant antitrust markets for merger analysis. The concept of a *relevant antitrust market* has a very specific meaning. In an antitrust analysis, one is specifically interested in determining sets of products and suppliers for which price increases or other anticompetitive actions would be profitable. At times during their analysis of the California utility merger, the staff of the Federal Energy Regulatory Commission confused antitrust markets with economic markets. The term *economic market* is frequently used to refer to a set of products and a geographic area for which prices tend toward uniformity, allowing for transportation costs. Since antitrust markets and economic markets are defined differently for distinct purposes, it should not be surprising that antitrust markets can be larger or smaller than economic markets (Morris and Mosteller 1991,

Scheffman and Spiller 1987; Werden 1981, 1983; Kimmel 1987; Baker 1987).

CONCLUSION

Market definition plays a key role in any analysis in which market power is at issue, and thus in all antitrust cases. The competitive analysis of proposed mergers often turns on what is included in relevant product and geographic markets. As a result, the definition of relevant markets is highly contentious and is often subject to abuse.

Following the *Merger Guidelines*, the federal antitrust agencies define markets using the hypothetical monopolist test. To determine whether a proposed group of products is a relevant market in which market power could be exercised, this test asks whether a hypothetical monopolist that was the sole supplier of those products would find it profitable to raise prices.

Market definition is not an end in itself but simply one step in the process of determining whether some group of consumers is likely to be injured through the exercise of market power. It is important to keep in mind the goal of the analysis, which should consider three questions in sequence: Which customers might be victimized by a price increase for Service X? Which other services are substitutes for Service X from the point of view of customers? Which suppliers can respond to an increase in the price of Service X by beginning to supply either Service X or substitutes? The rationale for this series of questions is that customers cannot be victimized if an increase in the price of Service X would not be profitable for suppliers. A price increase will not be profitable if enough customers can switch to another service without suffering significant adverse effects, or if enough customers can switch to a new supplier that would begin to supply Service X if it could sell its output.

NOTES

1. See, for example, *U.S. v. Country Lakes Foods*, CA 3-90-101 (D. Minn. 3rd Div. June 1, 1990), and *Williams Pipe Line Company*, Docket No. IS90-21-000 *et al.*, Initial Decision, 58 FERC ¶63,004, Jan. 24, 1992.

2. Terra Comfort Corp., *Amendment to Initial Filing*, "Analysis in Support of Market Based Rate," Federal Energy Regulatory Commission Docket ER90-40-000, 1990, Ex. D-1, 5–6.

3. Federal Energy Regulatory Commission Docket EC89-5-000, Southern California Edison and San Diego Gas and Electric, Exhibit 1677 (Paul L. Joskow),

Exhibit 1509 (Joe D. Pace), and Exhibit 1583 (William R. Hughes, stating that he, the administrative law judge, and Federal Energy Regulatory Commission all included trading in the transmission services market in the *Utah Power and Light* case).

4. Owen 1989, 120; Federal Energy Regulatory Commission 1990a, 59 TR 9341–42, 30 TR 4626–27, 33 TR 5021, 5110–11; Federal Energy Regulatory Commission, Docket EC89-5-000, Southern California Edison and San Diego Gas and Electric, Exhibit 1810 and Exhibits 1190 and 1191 (for San Diego Gas and Electric, on average incremental costs of oil and gas generation were 20 percent or more above the price of Southwest economy energy).

5. Federal Energy Regulatory Commission 1990d, 59–60. See also Southern California Edison and San Diego Gas and Electric 1990, n. 58, and Federal Energy Regulatory Commission Docket EC89-5-000, Southern California Edison and San Diego Gas and Electric, Exhibit 802, 78–82.

5

Competitive Effects of Horizontal Mergers

Antitrust analysis focuses on whether mergers will significantly increase the likelihood that market power will be exercised. This is the central concern of Section 7 of the Clayton Act. Plaintiffs do not have to prove that prices will increase.

In analyzing mergers, the Department of Justice and the Federal Trade Commission evaluate the likelihood that market power will be exercised by three types of behavior: (1) by a monopolist or by a dominant firm that competes with small "fringe" suppliers; (2) through unilateral, noncoordinated behavior in a market with a small number of suppliers; or (3) through collusion—that is, coordinated behavior—by a group of firms. Collusion may be either express or tacit. The antitrust agencies base their analyses on the premise that various structural characteristics of markets are associated with greater likelihood that market power will be exercised.

In contrast to the antitrust agencies, the Federal Energy Regulatory Commission tends to consider only a limited range of potential competitive theories. The Commission concentrates on the particular firm that is the object of its decision making, for example, the firm that would be created by a merger or the firm applying for authority to charge market-based prices (Tenenbaum and Henderson 1991). In the process, the Commission sometimes ignores the importance of the remaining structural elements of the market. This creates two types of problems. On the one hand, the Commission overlooks some of the competitive alternatives that constrain a firm's ability to exercise market power, making the Commission likely to perceive competitive problems where none exist. On the other hand, the Commission overlooks the possibility that market power might be exercised by the uncoordinated or coordinated actions of a group of firms, making it likely that it will overlook genuine competitive problems (Harris and Frankena 1992).

TYPES OF ANTICOMPETITIVE BEHAVIOR

If a merger would result in a monopoly, the remaining competitive issue is whether the exercise of market power would be rendered unlikely by ease of entry. If a merger would result in a dominant firm for which the only direct competitors are small fringe suppliers, an additional issue would be whether output from the competitive fringe would constrain the ability of the dominant firm to increase prices.

In most merger investigations, the federal antitrust agencies proceed on the theory that the exercise of market power would require collusion among a group of firms. In such cases, it is important to analyze not only market shares and entry conditions but also factors, including the concentration of firms in the market, that affect the ability of firms to reach and enforce a collusive agreement.

At one extreme is *express collusion* to reduce competition. Competitors might meet and reach an agreement on prices, outputs, sales territories, or advertising. Collusive agreements require policing because individual colluders have an incentive to expand sales if the price is above the competitive level. Colluders might exchange information or take collective actions to guard against cheating. Explicit collusion to reduce competition is, of course, unlawful, except in industries exempted from the Sherman Act.

Antitrust evaluation of mergers often focuses on *tacit collusion*—coordination of behavior to reduce competition in concentrated markets without an explicit agreement. By observing each other's behavior, firms in concentrated markets realize their interdependence and may form strategies to arrive at prices and profits above competitive levels without an express agreement (Tirole 1988, chap. 6; Jacquemin and Slade 1989). Merger evaluation under Section 7 of the Clayton Act is particularly concerned with avoiding market structures that would make tacit collusion likely, because detection and prosecution of tacit collusion are difficult.

Even if a merger leaves more than one substantial competitor in a relevant market, the federal antitrust agencies also evaluate the likelihood that the merger would lead to higher prices as a result of unilateral behavior, that is, decisions by individual firms that do not rely on the concurrence or coordinated responses of rivals. For example, suppose that among the firms supplying differentiated products in a relevant market, the products of the merging firms are the closest substitutes for each other. Alternatively, suppose that the merging firms are frequently the two bidders with the lowest prices. In such cases, an antitrust agency may conclude that, even without coordinated behavior, the merger would be likely to lead to higher prices (U.S. Department of Justice and Federal Trade Commission 1992, Section 2.21).

Another scenario involving unilateral behavior has been proposed for concentrated markets with entry barriers where firms recognize the effect of their output decisions on the market price. In deciding whether to increase its output level, a firm balances the margin it earns on incremental output against the reduction in margins on all its existing sales. The reduction in margins occurs because the incremental output depresses the market price. A merged firm will have a higher level of sales than either of the merging firms had before the merger. The merged firm, therefore, will lose more from reduced margins on existing sales than either merging firm alone would have lost if it had increased its output, and depressed the market price, by the same amount. As a result, the merged firm has less incentive to expand, and more incentive to reduce, output than either of the merging firms had before the merger. The net result could be less output and higher prices after the merger.

MEASUREMENT OF MARKET SHARES

The shares of firms in a relevant antitrust market are used along with other information to evaluate the likelihood of unilateral exercise of market power and of collusion by a group of firms. One must first determine how market shares should be measured so that they will reflect the competitive significance of the firms. For example, should one calculate each firm's share of output (production or sales), or each firm's share of capacity to supply the market? The *Merger Guidelines* (Section 1.41) indicate that the important issue is the magnitude of each firm's likely response to a "small but significant and nontransitory" increase in price (assuming the firms act competitively). Market shares should reflect the ability to thwart an anticompetitive price increase by expanding sales, and the ability to enforce a collusive agreement by threatening to punish cheaters by expanding sales and driving prices down.

Where the ratio of sales in a particular market to the capacity available to supply that market varies significantly among firms, the decision on how to measure market shares can influence measured concentration in a market and the increase in that concentration resulting from a merger. In the extreme case, a firm may play a competitive role in a market by virtue of having capacity available without actually making any sales.

Suppose Utility A can supply all units demanded, incurring a constant marginal cost of $20.00 per megawatt-hour. Utility B could supply all those units at a constant marginal cost to itself of $20.20 per megawatt-hour. There is no other source of supply at a marginal cost below $22.00 per megawatt-hour. Absent collusion, one would expect Utility A to make all the sales at a price between $20.00 and $20.20 per megawatt-hour, while Utility B would make no sales. Nonetheless, Utility B plays an important competitive role by constraining Utility A to charge less than $20.20 per megawatt-hour.

This argument was important in the antitrust evaluation of the proposed merger of Southern California Edison and San Diego Gas and Electric. In the case of short-term transmission from the Southwest to northern California, San Diego Gas and Electric's competitive role was understated by its actual sales. Its significance as an independent competitor arose from the fact that it offered an alternative that constrained the prices charged by Southern California Edison for short-term transmission service. Thus what was primarily pertinent was San Diego Gas and Electric's share of relevant capacity (Owen 1989, 146–47).

Measures of market share that reflect actual sales are typically used for differentiated products that sell on the basis of advertising, marketing, brand names, established customer relationships, after-sales service, and the like. Measures of market share that reflect capacity typically are used for homogeneous products that sell on the basis of specifications and price. Thus, if one were considering a retail energy market in which electricity competes with other sources of energy, market shares based on sales might be appropriate. On the other hand, one would expect that market shares based on capacity available to supply the market would be appropriate in analyzing bulk electric power markets.

In three situations, however, it would not be appropriate to base market shares on either total sales or total capacities. First, suppose that the marginal costs of *additional* output above the present level would be significantly higher for some firms than for others, even though the firms had similar sales and capacities. This could be the case if the unused generating capacity of some firms was at oil or gas facilities with a relatively high operating cost, while the excess capacity of other firms was at coal facilities with a significantly lower operating cost. In such a case, use of market shares based on sales or capacity could overstate the competitive roles of the firms with oil or gas facilities compared to those with coal facilities.

Second, some capacity may not be available for sales in a particular market. Suppose that the relevant market is short-term transmission services supplied to other utilities. Some of a firm's transmission capacity normally would not be available to supply that market, even if there were a "small but significant and nontransitory" increase in price. Utilities are required by regulation to use some capacity to supply power to their own retail customers. Only the capacity available to supply transmission services to others should be included in calculating shares in the relevant market.

Short-run transmission service was a relevant market in the proposed merger of Southern California Edison and San Diego Gas and Electric. Those firms argued that in the absence of information on the amount of capacity available to provide transmission service to others (information that they did not attempt to provide), market shares should be based on actual sales (Pace 1990, 34, 73–75). By contrast, some of the intervenors argued that in the absence of contrary information it was reasonable to assume that capacity to supply short-run transmission service to others was proportional to total capacity (Owen 1990b, 11–12). The California commission adopted the latter argument after a witness demonstrated that subtraction of capacity required to transmit energy under firm purchase agreements left concentration in the same range as that based on total capacity (California Public Utilities Commission 1991a, 39–40, and 1991b).

Third, differences in sales or capacity among firms may not be competitively significant. When this is true, it may be appropriate to assign firms equal market shares. Suppose that each firm has enough efficient capacity to satisfy all the demand for a homogeneous product or service. The fact that one firm has more capacity than another would have no competitive significance. This situation could arise if there were substantial excess capacity in a relevant transmission market.

Alternatively, suppose that a market involves bids for future delivery of a product or service, and that winning bidders have enough time to expand capacity and yet meet the contract delivery schedule. What is important is whether a firm can convince customers that it is a qualified bidder that can produce the product or service and meet the delivery schedule. Among the firms that are qualified bidders, the fact that one has had greater sales than another in the past may be irrelevant to the analysis of competition (Merger Standards Task Force 1986, 158–59). This situation is likely to arise with regard to long-term purchases of bulk power where customers request bids for future delivery (Harris and Frankena 1992).

CONCENTRATION

Concentration in a relevant market is measured in order to evaluate the likelihood that a merger would lead to or increase the exercise of market power. For an exercise of market power to be likely, the relevant market must be concentrated. As the *Merger Guidelines* specify, the federal antitrust agencies rely primarily on Herfindahl-Hirschman Indices (HHIs) to measure concentration (Section 1.5). The HHI is calculated by summing the squares of the market shares of the firms in the market. The smaller the number of firms and the more unequal their sizes, the larger the HHI will be, and by definition the more concentrated the market is. For example, if there are five equal-sized firms, each with 20 percent of the market, the HHI equals $5 \times (20)^2$ or 2,000. If there are four equal-sized firms, each with 25 percent of the market, the HHI equals $4 \times (25)^2$ or 2,500. If there are four firms that are not equal in size, the HHI will exceed 2,500. The highest possible HHI—where there is a monopolist—is 10,000.

The *Merger Guidelines* discuss the level and increase in concentration at which the federal antitrust agencies would be likely to challenge a horizontal merger. According to these guidelines, the agencies regard markets with a postmerger HHI between 1,000 and 1,800 to be moderately concentrated:

> Mergers producing an increase in the HHI of less than 100 points in moderately concentrated markets post-merger are unlikely to have adverse competitive consequences and ordinarily require no further analysis. Mergers producing an increase in the HHI of more than 100 points in moderately concentrated markets post-merger potentially raise significant competitive concerns depending on the factors set forth in Sections 2–5 of the Guidelines (Section 1.51.b).

Sections 2–5 of the *Merger Guidelines* deal with the likelihood of successful collusion or unilateral exercise of market power, ease of entry, efficiencies, and the failing firm defense.

If the HHI is above 1,800, under the *Merger Guidelines* the market is highly concentrated. The guidelines state:

> Mergers producing an increase in the HHI of more than 50 points in highly concentrated markets post-merger potentially raise significant competitive concerns, depending on the factors set forth in Sections 2–5 of the Guidelines. Where the post-merger HHI exceeds 1800, it will be presumed that mergers producing an increase in the HHI of more than 100 points are likely to create or enhance market power or facilitate its exercise. The presumption may be overcome by a showing that factors set forth in Sections 2–5 of the Guidelines make it unlikely that the merger will create or enhance market power or facili-

tate its exercise, in light of market concentration and market shares. (Section 1.51.c)

It is widely recognized that the HHI thresholds specified in the *Merger Guidelines* are not based on empirical evidence concerning the relationship between concentration and the likelihood that market power will be exercised (Pautler 1983; Uri and Coate 1987; Scherer and Ross 1990, chap. 11).

A study of Federal Trade Commission merger investigations and challenges during 1982–86 found that it was not uncommon for the Commission to investigate a merger where the postmerger HHI exceeded 1,800 and the increase in the HHI exceeded 100, but then decide not to challenge it because of the weight attached to entry conditions and other factors relating to the ability to collude. At the same time, for five of the twenty-seven mergers challenged by the Federal Trade Commission in the 1982–86 period, the postmerger HHI was between 1,000 and 1,800 (Coate and McChesney 1992). The latter could reflect unease with the market definitions used to compute the HHIs.

Throughout the past decade, there have been differences between what the various editions of the *Merger Guidelines* have stated about concentration thresholds, on the one hand, and the enforcement policies of the antitrust agencies, on the other (Scheffman 1993). Also, while the courts have used the analytical framework described in the *Merger Guidelines*, this is an agency policy statement that does not have the force of law (Newborn and Snider 1992).

Sometimes it is impossible to obtain all the data needed to calculate the HHI for a market. When that is the case, two shortcuts can be useful. First, if the number of firms in the market is known but their market shares are not known, one can determine a lower bound on the HHI by dividing the number 10,000 by the number of firms, *n*. When there are *n* equal-sized firms in the market, the HHI is (10,000/*n*). The more unequal those firms are in size, the more the HHI will exceed (10,000/*n*). Thus, whenever there are five or fewer firms in the market, the HHI will be at least 2,000.

Second, the increase in the HHI that results from a merger can be computed by multiplying the market shares of the two merging firms together and then multiplying the product by 2. For example, if firms with market shares of 7 percent and 5 percent merge, the increase in the HHI is 70. Thus, it is possible to compute the increase in the HHI without knowing the shares of firms other than the ones that are merging.

Putting these two shortcuts together, it is sometimes possible to determine from incomplete data that the postmerger HHI exceeds 1,800 and that the increase in the HHI as a result of the merger is at least 100. For example, suppose that in a market with six firms, two firms, each with a share of at least 10 percent, propose to merge. Without knowing more, one can conclude that the postmerger HHI would be at least 2,000 (10,000/5) and that the increase in the HHI would be at least 200 (10 x 10 x 2).

ENTRY CONDITIONS

Entry conditions are an important structural factor to consider in determining whether a merger increases the likelihood that market power will be exercised. As the *Merger Guidelines* state: "A merger is not likely to create or enhance market power or to facilitate its exercise, if entry into the market is so easy that market participants, after the merger, either collectively or unilaterally could not profitably maintain a price increase above premerger levels" (Section 3.0). Accordingly, the antitrust agencies do not challenge mergers when they find that entry is easy. However, the antitrust agencies have been known to find that entry is not easy when the evidence is to the contrary. The standards recently applied by the courts in accepting ease of entry as a defense for mergers in concentrated markets have been less stringent than those applied by the Department of Justice. In several recent merger cases brought by the Department, federal appeals courts decided in favor of defendants on the grounds that entry or its threat would prevent the exercise of market power (*United States v. Waste Management*, 743 F.2d 976 (2d Cir. 1984); *United States v. Syufy Enterprises*, 903 F.2d 659 (9th Cir. 1990); *United States v. Baker Hughes, Inc.*, 908 F.2d 981 (D.C. Cir. 1990). See also *United States v. Calmar Inc. and Realex Corp.*, 1985-1 Trade Cas. (CCH) ¶66,588 (D.N.J. 1985)).

The *Merger Guidelines* state that entry is "easy" if entry is *timely*, *likely*, and *sufficient* in magnitude to deter or counteract potential anticompetitive effects from a merger (Section 3). For timeliness, the guidelines use a standard of two years between initial planning and "significant market impact." For likelihood, the guidelines use a profitability standard. Postmerger entry that would be profitable at premerger prices is considered likely. If entry would not be profitable at premerger prices, then it is considered unlikely. For sufficiency, the guidelines use as a standard whether the entry would return prices to premerger levels.

Given the two-year timeliness threshold, entry is likely to be considered difficult in many markets relevant to mergers in the electric utility industry. Major transmission and generating projects take about five to eight years from initial planning to operation (Owen 1989, 131–34). They face serious opposition from environmentalists. Hence there are likely to be many situations in which impediments to entry into short- and medium-term bulk power and transmission markets would be significant under the *Merger Guidelines*. In light of state laws and regulations relating to the territories of distribution companies, barriers to entry in retail distribution of electricity would generally be high.

An exception would be generation in situations where cost-effective smaller units can be installed in less than two years. Generating facilities intended to provide capacity without supplying much associated energy may be an example. In a case recently reviewed by the Federal Energy Regulatory Commission, the firm that won a contract to supply delivered bulk power planned to provide the capacity—apparently in well under two years—using secondhand, relocated combustion turbines (Harris and Frankena 1992).

There would also be situations involving long-term contracts for future delivery in which entry may be relatively easy. Delivery on a contract may not have to begin sooner than the necessary facilities could be planned, constructed, and brought into operation. For example, in a matter recently reviewed by the Federal Energy Regulatory Commission, the firm that won a contract to supply delivered bulk power planned to build a new transmission line to fulfill the contract (Harris and Frankena 1992). Because it could obtain a contract before making its investment, the firm substantially reduced the risks that might otherwise have been involved in entry because of sunk costs. Such entry can be facilitated by customers that would otherwise be subjected to an anticompetitive price increase. Of course, this assumes that no other problem limits entry. For example, it assumes that ownership of potential transmission corridors or generating sites is not sufficiently concentrated to be a problem and that approvals to build new facilities could be obtained.

In the case of long-term supplies, an important potential entrant may be the customer. As an alternative to facilitating the entry of another firm by entering into a contract, the customer may be in a position to integrate vertically. The issue is whether there would be a significant cost penalty associated with such vertical integration. Backward vertical integration by a buyer (or entry facilitated by contracts with a buyer) would be most likely to constrain the exercise of market power where the requirements of one buyer or a few buyers are large relative to the minimum efficient scale for entry.

ABILITY OF FIRMS TO COLLUDE

In evaluating mergers the federal antitrust agencies consider the ability of firms to collude (*Merger Guidelines,* Section 2). To coordinate their pricing and output successfully, firms must be able to arrive at a mutually beneficial agreement, and they must be able to detect and retaliate against cheaters. Among the factors suggested by the *Merger Guidelines* or advanced in litigation as determinants of the ease of collusion are: availability of information that would facilitate detection of cheaters, product heterogeneity, volatility of demand, differences in costs among suppliers, past collusion in the market, the existence of sophisticated large buyers, and the frequency and size of orders. With respect to the last point, the guidelines state:

> In certain circumstances, buyer characteristics and the nature of the procurement process may affect the incentives to deviate from terms of coordination. Buyer size alone is not the determining characteristic. Where large buyers likely would engage in long-term contracting, so that the sales covered by such contracts can be large relative to the total output of a firm in the market, firms may have the incentive to deviate. However, this only can be accomplished where the duration, volume and profitability of the business covered by such contracts are sufficiently large as to make deviation more profitable in the long term than honoring the terms of coordination, and buyers likely would switch suppliers. (Section 2.12)

Accordingly, when orders for the relevant product are relatively infrequent and large, the exercise of market power is less likely. Examples can be drawn from competitive bids for long-term contracts involving aerospace and weapons systems, telecommunications services, and electric power (Harris and Frankena 1992). If contracts are awarded to a single firm, or are split in such a way that the first-place firm gets a significantly larger contract than the second-place firm, each bidder has a strong incentive to win. Absent side payments (that is, bribes from winners to losers) and absent future bids in which first-round losers would have a turn to win, there would be no way for firms that lost the initial bid to share the profits from prices above competitive levels. Similarly, there would be no way to retaliate against first-round cheaters and hence no way to deter cheating. In such cases one might not be concerned about collusion even with relatively few firms.

There are thus two related arguments for why the exercise of market power may be unlikely when there are large buyers in a market. First, if the purchases of a small number of buyers are large relative to the minimum viable scale for a supplier, those buyers may be able to integrate vertically or to use contracts to facilitate new entry. Second, if the purchases of a small number of buyers are large relative to the sales of a supplier whose cooperation is necessary for the success of a collusive price increase, these buyers may be able to induce that supplier to cheat on the collusive agreement. A supplier would be likely to cheat if it would profit more from its markup on additional sales than from an anticompetitive price increase on the lower volume of sales it would make if it participated in the exercise of market power. These two arguments have played a role in a number of recent merger cases.

REGULATION

The electric utility industry is regulated at the federal and state levels. Nonetheless, a utility merger involving horizontal overlaps, high concentration, and entry barriers can have adverse competitive effects. In examining the role of regulation in mitigating anticompetitive effects, it is useful to ask two questions. First, given regulation, how could the merged firm (alone or in collusion with others) raise its prices or otherwise exercise market power where competition would be reduced by a merger? Second, why would a firm subject to rate-of-return regulation have an incentive to raise its prices or otherwise exercise market power in some markets?

Ability to Exercise Market Power

There are three reasons why a regulated firm would have the ability to exercise market power if competition were to decline as a result of a merger. First, regulation may, in effect, set ceiling prices that are above competitive rates. Regulated rates typically are set to permit recovery of fixed costs and what is deemed a fair rate of return. Competition among suppliers with excess capacity would lead to rates based on short-run marginal costs that do not cover fixed costs. Thus a reduction in competition could permit an increase in rates, even if the rates could not be raised all the way to the monopoly level.

Second, regulation is incomplete and in some circumstances can be evaded. Regulation of prices for bulk power in coordination transactions among utilities is loose. In some situations, regulated firms with market power in distribution or transmission may be able to charge rates above competitive levels by paying inflated prices to unregulated generation affiliates.

Third, because regulation has been responsible for wasting resources, the extent of regulation of the electric utility industry is being reduced.

Increased reliance is being placed on market forces. It is important to preserve competition in areas that are likely to be deregulated in the future. Acceptance of an increase in concentration on the grounds that the industry is now regulated would reduce the future scope for deregulation and hence benefits from increased competition.

Incentive to Exercise Market Power

The extent to which shareholders can profit from the exercise of market power varies. For investor-owned utilities in California, the increased profits associated with a higher markup on sales and resales of power are shared by shareholders and ratepayers; shareholders receive 8 percent (Federal Energy Regulatory Commission 1990a, 33 Tr. 5137, 34 Tr. 5170; Mays 1990, 46, 72; Boettcher 1990, 68). Public Service of Oklahoma recently proposed that shareholders' share of profits from off-system sales be increased from 10 to 25 percent "to ensure that efforts are continued to maximize profits from off-system electric sales" (*Electric Utility Week,* Oct. 26, 1992, 9). Once its rate of return reaches 10.8 percent, Public Service of Indiana retains 30 to 90 percent of additional ex-post profits, subject to earning a maximum rate of return of 12.3 percent (*Electric Utility Week*, Nov. 16, 1992, 11).

Even with a binding constraint on the rate of return, an increase in prices and profits in one market enables a regulated firm to lower prices in other markets. A price reduction in another market allows the firm to increase its rate base and, if the allowed rate of return exceeds the cost of capital, thereby increase its total profits. In addition, after raising prices, the managers could allow costs to rise, either by increasing their own salaries and benefits or by reducing their efforts. It is also possible for a regulated firm to exercise market power in ways that increase the profits of unregulated affiliates.

HORIZONTAL EFFECTS ON COMPETITION

The next step in an antitrust analysis is to reach a conclusion on whether a horizontal merger is likely to lead to a significant reduction in competition in any relevant market. There is some confusion regarding the standards for antitrust liability in connection with mergers that arises from the *Merger Guidelines*' specification of a hypothetical "small but significant" or 5 percent price increase in defining relevant markets and identifying uncommitted entrants. People evaluating proposed mergers have wrongly inferred that the federal antitrust agencies tolerate price increases of up to 5 percent resulting from a merger (Baughcum 1989, 14–15; Pace 1990, 60). The *Merger Guidelines* explain: "The 'small but significant and non-transitory' increase in price is employed solely as a methodological tool for the analysis of mergers: it is not a tolerance level for price increases" (Section 1.0).

Typically there is uncertainty about the likelihood that a proposed merger will reduce competition. The conclusion reached by the antitrust agencies regarding competitive effects may, therefore, depend on whether customers in a relevant market—that is, the potential victims of an exercise of market power—oppose or support the merger. Of course, customers may oppose a merger that is competitively benign. The customers may misunderstand the potential for market power to be exercised, or they may oppose the merger in the hope of exacting a beneficial settlement. The conclusion reached by the antitrust agencies may also depend on information and opinions they find in company documents bearing on competitive effects.

MONOPSONY POWER

The preceding discussion has concentrated on monopoly power of sellers. The same methodology can be applied to mergers that might increase monopsony power of buyers. According to the *Merger Guidelines*:

> Market power also encompasses the ability of a single buyer (a "monopsonist"), a coordinating group of buyers, or a single buyer, not a monopsonist, to depress the price paid for a product to a level that is below the competitive price and thereby depress output. The exercise of market power by buyers ("monopsony power") has adverse effects comparable to those associated with the exercise of market power by sellers. In order to assess potential monopsony concerns, the Agency will apply an analytical framework analogous to the framework of these Guidelines. (Section 0.1)

The federal antitrust agencies have challenged mergers on the grounds that they increased the likelihood of exercise of monopsony power. For example, when InterNorth acquired Houston Natural Gas, the Federal Trade Commission alleged a relevant market for the purchase of natural gas in certain producing fields and basins and the transportation of that gas to consumers. The Federal Trade Commission's complaint was settled by a consent order requiring the divestiture of a number of pipeline interests (*InterNorth et al.*, 106 FTC 312 (1985)).

In the Southern California Edison/San Diego Gas and Electric merger case, California regulators determined that the merged entity would be able to exercise buyer market power, particularly for bulk economy energy purchased in the Southwest (California Public Utilities Commission 1991a, 779, 810, 835). The relevant market was defined as bulk economy energy purchases in Texas, Colorado, Arizona, New Mexico, southern Nevada, Utah, and southern California. The Division of Ratepayer Advocates of the California Public Utilities Commission estimated that the HHI would increase by 350 to 700 and that the postmerger HHI would be between 1,140 and 1,740. The California commission rejected a recommendation by one of the applicants' economic experts that weaker structural standards should apply to buyer market power than those used by the Department of Justice in connection with seller market power. In addition, the California commission concluded that "the proposed merger will adversely impact competition in the SW nonfirm bulk power market by enabling the merged entity to exercise buyer market power" and that it "may" do so in the Pacific Northwest (California Public Utilities Commission 1991b, 61–64).

CONCLUSION

Once relevant antitrust markets have been defined, a relatively standard methodology is used by the antitrust agencies to determine the likely competitive effects of a horizontal merger. The first step is to calculate market shares so that they, and measures of concentration based on them, accurately reflect the competitive significance of a transaction. The second step is to calculate market concentration using the Herfindahl-Hirschman Index (HHI), premerger and postmerger. Sufficiently high postmerger concentration levels and increases in concentration establish a rebuttable presumption of anticompetitive effects. Persuasive evidence that entry is easy is sufficient to rebut the presumption of anticompetitive effects. In the electric utility industry, however, entry is likely to be difficult in many situations. As a result, market definition and the manner in which market shares are calculated are particularly important. The presumption also may be rebutted by showing that the postmerger firm would not have unilateral ability or incentive to exercise increased market power and that collusion would be unlikely in the market.

Before a decision can be made to challenge a merger under the antitrust laws, or under regulatory statutes based on competitive effects, three more actions should occur: evaluation of the likelihood of anticompetitive effects because of vertical aspects of the proposed merger (Chapter 6); consideration of remedies for likely competitive problems (Chapter 8); and evaluation of efficiencies that would result from the merger (Chapter 9).

6

Competitive Effects of Vertical Mergers

Rate regulation of firms with market power, including local distribution companies in the electric power industry, is intended to constrain the exercise of market power. Regulation forces the firm to charge a price below the monopoly price. Antitrust concerns may arise when a firm that has a regulated monopoly engages in another activity in which it faces competition. Such a combination may allow the firm to evade the regulatory constraint on its exercise of market power.[1] For example, by allocating costs for unregulated competitive activities to the regulated activity, the firm may obtain regulatory approval for an increase in cost-based prices for the regulated activity. Such a price increase involves an exercise of market power. At the same time, the firm realizes increased profits. While the increased profits actually come from the regulated activity, from an accounting point of view they accrue to the unregulated competitive activity.

Greater antitrust problems occur when there is a vertical relationship between the regulated monopoly and the unregulated competitive activity. Two entities are vertically related when one supplies a product that is used by the other as an input in production. Suppose a firm has a regulated monopoly in a downstream activity (for example, retail distribution of electricity), and the firm also owns upstream facilities that are unregulated (for example, electrical generating plants). The vertical relationship affords the firm opportunities to exercise increased market power in distribution and realize the resulting higher profits from the unregulated activity. Put simply, the unregulated generating facility may be able to sell power at inflated prices to the distribution company, which then is allowed to raise prices to retail customers.

The exercise of additional market power in the distribution market—that is, *evasion of rate regulation*—is not the only antitrust problem that arises in this situation. The other antitrust problem is *foreclosure of competition* from lower cost producers. The monopoly distribution company has an incentive to buy power from its unregulated generating facility even if the latter is not the lowest-cost producer available if the higher costs can be passed along to retail customers in higher regulated prices.

These two competitive problems—evasion of rate regulation and foreclosure of competition—arise in the electric power industry in two instances: (1) where there is common ownership of local distribution companies and electric generating facilities that are unregulated independent power producers or that are qualifying facilities under the terms of the Public Utility Regulatory Policy Act of 1978, and (2) where there is common ownership of transmission and generating facilities. In the latter case, the exercise of market power may involve denial of transmission service to competing producers of electric power. A utility with market power in transmission may deny transmission service to a competing producer of electric power in order to protect its own bulk power sales. The incentive to do this is likely to be greatest when the utility with the transmission facilities also has an ownership interest in non-rate-based generating capacity.[2]

The problematic vertical integration at issue here may arise or increase in extent as a result of a merger, in which case the merger raises vertical competitive issues. Of course, common ownership of vertically related units can arise not only through mergers but also through internal growth and diversification of a firm. In connection with the proposed merger between Southern California Edison and San Diego Gas and Electric, we argued—and regulators agreed—that the extent of the problematic common ownership would be greater with the merger than without it. Regulators also found that vertical aspects of the merger between Utah Power and Light and Pacific Power and Light, as well as the merger between Northeast Utilities and Public Service of New Hampshire, raised competitive concerns.

THE VERTICAL THEORY: FOUR CONDITIONS

A vertical merger may facilitate anticompetitive evasion of regulation by a firm paying inflated prices for inputs purchased from an upstream affiliate, and it may encourage foreclosure of upstream competition (Morris 1992). Four conditions are sufficient to raise concerns that a vertical merger would have such anticompetitive effects (Owen 1993a, 1993b).

1. *Market Power.* The first condition is that the downstream unit has market power. Absent regulation, it could increase its profits by raising prices above the competitive level. Whether a firm has market power depends on the existence of competing suppliers of the same products, the existence of substitute products, and the ease of new entry.

2. *Regulation.* The second condition is that the downstream unit is subject to cost-based regulation that constrains prices to be below the level that would maximize the firm's profits. In other words, the downstream unit has unexercised market power as a result of regulation.

3. *Upstream Activity.* In the situation of primary concern in this chapter, the upstream unit is unregulated and faces competition. Because it is unregulated, it can realize increased profits that are transferred to it through inflated prices paid by the downstream unit. Because it faces competition, its expansion reduces the market shares of rivals, including ones that may have lower costs.[3]

4. *Imperfect Regulation.* The fourth condition is that the merged firm can reasonably expect to obtain approval for an increase in its regulated prices based on the higher costs that result from paying inflated prices for inputs to the affiliated unit. In other words, the regulatory agency is not able to monitor transfer prices perfectly. Similarly, under variations in the theory, the regulatory agency cannot monitor perfectly various other activities, such as service denials, the nonprice terms of contracts, and cost accounting by the merged firm. Perfect monitoring of only some terms at which the monopoly product is sold might lead to changes in other terms in order to evade regulation. For example, effective regulation of transmission rates could lead to denial of transmission service to competing producers of electric power.

In principle, the incentives for attempted evasion of regulation that are created by imperfect monitoring might be offset by appropriate penalties exceeding mere disallowances of overcharges and disgorgement of ill-gotten gains. If penalties are sufficiently high, a firm's expected profits from attempts to evade regulation could be negative even if the likelihood of being caught on any specific attempt is less than 100 percent. Thus, in principle, the existence of penalties high enough to deter attempts to evade regulation could mitigate problems relating to mispricing in affiliate transactions.

Anyone familiar with regulation of affiliate transactions is aware of the difficulty and the cost of attempting to rely on regulatory oversight to avoid abuses. It is impossible for regulators, with their limited resources and numerous responsibilities, to evaluate accurately all relevant purchase contracts for electric capacity and energy—particularly given the many variables that could affect prices (for example, type of capacity and energy, contract date, contract duration, location, and risk allocation). Regulators themselves agree. For example, the staff of the California Public Utilities Commission's Division of Ratepayer Advocates described the task of "auditing and assessing the reasonableness for ratepayers of transactions between utilities and their corporate relations" as "monumentally difficult, painful, controversial and litigious," and it stated that it "is not equipped to monitor and investigate all possible abuses in utility transactions with affiliates and headquarters" (California Public Utilities Commission, Division of Ratepayer Advocates 1988, IV-26; see also Ahern 1985, i, 11–15, 21).

A company can do a number of things to hamper evaluation of contracts by regulators. Suppose the regulator attempts to compare contracts between the regulated firm and affiliated companies to contracts between the regulated firm and nonaffiliated companies. The integrated regulated firm could inflate the prices paid in nonaffiliate transactions to avoid detection of inflated prices in affiliate transactions. The higher prices paid to nonaffiliates would facilitate evasion of regulation and could be passed along to ratepayers. The integrated regulated firm also could deliberately make its contracts with affiliates differ in many ways from its contracts with nonaffiliates so that the contracts could not effectively be compared.

Regulators monitoring the behavior of firms in complex industries, such as electric power and telecommunications, depend heavily on information from the firms they regulate. In the California commission's review of Southern California Edison's nonstandard contract with its affiliated qualifying facility, Kern River Cogeneration Company (KRCC), the Division of Ratepayer Advocates found that "Edison and KRCC made many efforts to thwart full discovery of facts by Commission staff, and to prevent the public disclosure of key facts" (Murray 1989). The administrative law judge involved in this matter criticized Edison's strategy of waiting until the rebuttal stage to submit evidence that should have been presented at the initial direct evidence submittal stage. Indeed, Edison's responses during discovery were found to be woefully inadequate (Myers 1989, 2366–75).

Similarly, in its petition for dismissal of an application by Edison to the California commission for approval of a transmission line, the Division of Ratepayer Advocates cited stonewalling and deception by Edison: "Edison has intentionally kept secret from the Commission staff critical agreements concerning this proposed project thereby misleading the Commission and public regarding the objectives and benefits of the project" (California Public Utilities Commission, Division of Ratepayer Advocates 1987, 1). In 1991 the California commission found that Edison had failed to meet its standards for providing information about nonstandard contracts with affiliated qualifying facilities (California Public Utilities Commission 1991c).

Regulation of affiliate transactions has not been very successful in preventing the abuses we have been discussing; it also has been enormously costly to taxpayers. The cost of trying to prevent abuses in affiliate transactions must be considered as a cost of a proposed vertical merger of the type we have described.

VERTICAL THEORY IN ANTITRUST ENFORCEMENT

The evasion-of-regulation and foreclosure-of-competition theory is explicitly stated in the U.S. Department of Justice's 1984 *Merger Guidelines* as an antitrust policy concern that will be taken into consideration in the evaluation of vertical mergers.[4]

> 4.23 Evasion of Rate Regulation
>
> Non-horizontal mergers may be used by monopoly public utilities subject to rate regulation as a tool for circumventing that regulation. The clearest example is the acquisition by a regulated utility of a supplier of its fixed or variable inputs. After the merger, the utility would be selling to itself and might be able arbitrarily to inflate the prices of internal transactions. Regulators may have great difficulty in policing these practices, particularly if there is no independent market for the product (or service) purchased from the affiliate. (A less severe, but nevertheless serious, problem can arise when a regulated utility acquires a firm that is not vertically related. The use of common facilities and managers may create an insoluble cost allocation problem and provide the opportunity to charge utility customers for non-utility costs, consequently distorting resource allocation in the adjacent as well as the regulated market.) As a result, inflated prices could be passed along to consumers as "legitimate" costs. . . . The Department will consider challenging mergers that create substantial opportunities for such abuses. (U.S. Department of Justice 1984b)

This vertical theory played an important role in the Justice Department's suit that led to the breakup of AT&T (Noll and Owen 1988, 1994; Brennan 1987, 1990). Similarly, in the case of natural gas, the Federal Trade Commission relied on the evasion-of-regulation theory in its complaint against the merger of Occidental Petroleum and MidCon, and action by the Federal Trade Commission based on this theory led to the divestiture of the Mississippi River Transmission pipeline.[5]

Ronald Carr, then Deputy Assistant Attorney General, explained the Antitrust Division's enforcement policy in the context of natural gas mergers:

> [T]he Division will scrutinize with care acquisitions of production companies and gas reserves by natural gas pipelines. Where we find that a pipeline has market power and is likely to be able to circumvent rate regulation by passing excessive gas prices through to consumers, the Division will challenge the acquisition.
>
> The Division is also concerned with acquisitions by regulated companies in the industry of unregulated companies not in the natural gas industry.
>
> By diversifying, a regulated entity will acquire the opportunity to allocate joint and common costs properly associated with its non-regulated enterprises to its regulated business and in this way circumvent rate regulation of the regulated business.
>
> Our experience in the *AT&T* case has convinced us that regulators are unlikely to be able to police such activities adequately, and we see no reason to be less concerned about this problem in the natural gas industry. (Carr 1982, 4, 9, 11–12)

In *United States v. AT&T,* the government's antitrust suit charged that through inflated prices between vertical affiliates, cross-subsidization, and discrimination against rivals, the Bell System had violated the antitrust laws by charging anticompetitive service and equipment prices to the customers of its regulated local telephone services, and by using its status as a regulated monopoly in local telephone service to foreclose competition in and monopolize the potentially competitive long-distance telephone service and telephone equipment markets. AT&T was vertically integrated into unregulated manufacturing of telephone equipment through its ownership of Western Electric.

In its antitrust suit, the government called as witnesses several former officials of the Federal Communications Commission who testified to the ineffectiveness of regulation in preventing abuses. The government's view was that the Bell System could not be regulated adequately to prevent abuses if regulated monopoly and competitive activities were combined. The Justice Department stated at the time that:

> At the heart of the government's case in *United States v. AT&T* was the failure of regulation to safeguard competition in the face of the powerful incentives and abilities of a firm engaged in the provision of both regulated monopoly and competitive services. Neither of these problems [cross-subsidization and discrimination] has thus far proven amenable to successful regulatory solution. Indeed, the very basis for divestiture is that the anticompetitive problems inherent in the joint provision of regulated monopoly and competitive services are otherwise insoluble.[6]

The settlement of the *AT&T* case provided for separate ownership for local telephone service, on the one hand, and long-distance service, equipment manufacturing, and information services, on the other. One of the government's operating premises in the antitrust case was that long-distance service could be competitive and hence eventually could be deregulated. Therefore, the settlement allowed AT&T to continue to operate regulated long-distance service and unregulated equipment manufacture. As a step in the direction of deregulation, the Federal Communications Commission has moved away from rate-of-return regulation for long-distance service.

In a matter involving attempts to remove line of business restrictions imposed on the regional telephone companies by the consent decree in *United States v. AT&T*, the U.S. District Court stated that regulators were indeed unable effectively to monitor the behavior of telephone companies:

> [T]he Court concluded following the close of the Department's case, and in accordance with the arguments presented by the Department, that "the Commission is not and never has been capable of effective enforcement of the laws governing AT&T's behavior," and that accordingly AT&T had been able to violate the antitrust laws in a number of ways over a long period of time with respect to interexchange [long-distance] services and the procurement of equipment. (*United States v. Western Electric Co.*, 673 F. Supp. 525, 531 (D.D.C. 1987), footnotes omitted)

Furthermore, the Court upheld the validity of the economic theory originally put forward by the Department of Justice:

> The Court found that, were the Regional Companies allowed to be present in the interexchange market while they also maintained monopoly control of the local telephone markets, they would be able to pursue precisely the same course as had the Bell System: (1) to discriminate in a variety of ways against their non-monopoly competitors through judicious use of the local monopoly; and (2) to "subsidize their interexchange prices with profits earned from their monopoly services." (673 F. Supp. 542)

The Court elaborated as follows on how the Bell System had foreclosed competition in unregulated markets:

> The manufacturing restriction was based in substantial part on evidence presented by the Department of Justice at the trial of this case indicating that the Bell System had improperly monopolized the market for telecommunications equipment, in that its local Operating Companies purchased such equipment primarily from Western Electric Company, the System's manufacturing

affiliate, and "engaged in systematic efforts to disadvantage outside suppliers." (673 F. Supp. 552)

As testimony and other evidence demonstrated, the Operating Companies managed, by one stratagem or other, to purchase Western Electric's products, even when those products were more expensive or of lesser quality than alternative goods available from unaffiliated vendors. (673 F. Supp. 553)

The Bell System subsidized the prices of its equipment with the revenues from the Operating Companies' monopoly services. The effect of this practice, as with respect to cross-subsidization generally, was (1) to permit the Bell System to undercut other producers of equipment (which lacked such a subsidy), and (2) unfairly to burden the consumers with excessive rates for the monopoly services they were furnished by the Operating Companies. (673 F. Supp. 553–54)

Moreover, the Court noted that:

Under the law, serious competitive concerns are raised even when relatively small market shares, for example as low as seven or eight percent, would be foreclosed as a result of leveraging of regulated monopolies into a related but unregulated market. . . . This leveraging doctrine serves antitrust interests by assuring that more efficient producers are not excluded from the market, and it prevents frustration of public regulation of subscriber rates. (673 F. Supp. 542, 556, 566–67)

In 1991, as a result of action by the U.S. Court of Appeals for the District of Columbia and against his own judgment, Judge Greene issued an order lifting the information service restrictions imposed by the AT&T consent decree on the regulated regional Bell operating companies.

Since 1987 the Antitrust Division of the Department of Justice has argued in favor of allowing the Bell operating companies (BOCs) to integrate vertically into unregulated equipment manufacturing.

The concern that a BOC might purchase inferior equipment from or pay excessive prices to its manufacturing affiliate at ratepayer expense is alleviated by regulations governing affiliate transactions and by changes in the marketplace—particularly the divestiture and the ability of regulators to make benchmark comparisons of BOC purchasing activities. Under current regulation, federal and state regulators can scrutinize and disallow excessive equipment costs and, if necessary, could impose additional restrictions on BOC self-dealing. Moreover, . . . strengthened Federal Communications Commission rules governing cost accounting and allocation alleviate the concern that the BOCs will engage in anticompetitive cross-subsidization of unregulated activities with ratepayer revenue. (Rill 1991, 5)

An additional rationale offered in the past by the Antitrust Division appears to be that unless foreclosure involves harm to third-party customers, it is a regulatory, not an antitrust, problem:

When, as a result of regulatory evasion, the only harm is to consumers in the bottleneck market—whom I will call the ratepayers—the appropriate antitrust response is inaction. This is not because ratepayers are unimportant, but rather because antitrust intervention—really antitrust interference—is likely to be counterproductive. An antitrust action designed to protect the very consumers that the regulation is designed to protect will almost inevitably make policy coordination difficult, diffuse responsibility, and threaten the success of efforts to prevent evasion. (Rule 1988, 11)[7]

The about-face by the Justice Department appears to have been due principally to a jurisdictional judgment—namely, that regulators, not antitrust courts, should determine whether conditions for anticompetitive vertical integration (or other integration) have been met. Judge Greene's increasingly detailed regulation of the telephone industry had made the Justice Department uncomfortable.

However, in the case of electric utility mergers, regulators seem to have accepted both the underlying vertical theory and the implications regarding their own limitations. State and federal regulators evaluating the competitive effects of the Utah Power and Light/Pacific Power and Light, Southern California Edison/San Diego Gas and Electric, and Northeast Utilities/Public Service of New Hampshire mergers based their findings concerning anticompetitive effects in part on the evasion-of-regulation and foreclosure-of-competition theory. Thus the economic theories that led the Government to seek the dissolution of AT&T are alive and well and being applied actively to electric utility mergers.

Two distinctions between the *AT&T* case and the electric utility mergers are worth noting. *United States v. AT&T* was litigated in court; the utility mergers were evaluated by regulators. And *United States v. AT&T* was litigated under Section 2 of the Sherman Act, while the utility mergers were decided under a weaker incipiency standard for anticompetitive effects based on Section 7 of the Clayton Act.

ABUSES IN EDISON'S AFFILIATE PURCHASES

Bulk power purchases by Southern California Edison from its sister company, Mission Energy, are an example of how a firm that owns a regulated monopoly and an unregulated competitive supplier of inputs to that monopoly may try to evade rate regulation and to foreclose competition.[8] Edison operates a local distribution company that is subject to state rate-of-return regulation. Mission Energy has a 50 percent interest in numerous unregulated generating facilities, including the Kern River Cogeneration Company, that are qualifying facilities under the Public Utility Regulatory Policy Act.

The California Division of Ratepayer Advocates reviewed the reasonableness of Edison's power purchases from Kern River during 1984–87 and concluded that Edison engaged in precisely the behavior contemplated by the evasion-of-regulation theory. To increase the profits of an unregulated subsidiary, Edison purchased power at prices higher than its avoided costs, and it attempted to pass on the higher costs to ratepayers in its regulated rates. Nonprice contract terms that shifted risks to ratepayers also were found to be unreasonable (California Public Utilities Commission, Division of Ratepayer Advocates 1988, chap. 1, and 1989).

The Division of Ratepayer Advocates found that abuses relating to the Kern River contract were not limited to the original terms of the contract but included the contract's administration and amendment:

> Edison paid the affiliated QF venture $3 million in bonus payments under the contract, but to "qualify" KRCC for the bonus, Edison let it reclassify a forced unscheduled outage to "scheduled maintenance." Monitoring Edison's contract administration with affiliated QF's over the long term will, in view of this conduct, be impossible to conduct effectively.
>
> KRCC and Edison "amended" the contract in 1988—in the face of DRA's investigation—so it would provide a firmer obligation to deliver power, more in line with the high capacity prices. Had any non-affiliated QF tried in 1988, an era of overcapacity for Edison, to convert an "as available" contract to "firm," it would have been summarily refused. (Murray 1989)

The division concluded that Edison's abuses could not adequately be handled by regulatory scrutiny of the contracts between Edison and its affiliated qualifying facilities. Divestiture was the preferred remedy. Failing divestiture, the division recommended that Edison not be allowed to purchase power from its affiliated qualifying facilities, or that the profits of the affiliated qualifying facilities from which it purchased be regulated (California Public Utilities Commission, Division of Ratepayer Advocates 1988, ¶25).

The California commission spent substantial resources reviewing the Kern River contract for the 1984–87 period. The commission agreed with the Division of Ratepayer Advocates "that the result of all of the KRCC contract terms was to create an agreement for as-available as opposed to firm capacity," while the prices paid by Edison were based on the avoided cost of firm capacity. The commission also found the division's concerns with the existence of self-dealing relating to the

Kern River contract to be legitimate. The commission ordered that Edison could not charge its ratepayers for $37.5 million (plus interest) that it paid to Kern River. The commission decided against prohibiting Edison from owning qualifying facilities, as the division had proposed. It also allowed Edison to sign "standard offer" power purchase contracts with affiliated suppliers without prior commission approval.[9] But if Edison wished to sign a nonstandard contract to purchase power from a supplier in which it had a financial interest, it was required to seek prior approval from the commission (California Public Utilities Commission 1990, 116–40, 154–58, 165–69).

REJECTION OF THE CALIFORNIA MERGER

Southern California Edison has a history of extensive power purchases from unregulated qualifying facilities in which its parent company, SCEcorp, has a 50 percent ownership interest through its Mission Energy subsidiary. As we have discussed, Edison engaged in various abuses designed to evade rate regulation in connection with its purchases of power from an affiliated qualifying facility, Kern River Cogeneration Company, during 1984–87. San Diego Gas and Electric, another investor-owned utility, had no similar affiliated qualifying facilities. In 1988 Edison and San Diego Gas and Electric reached an agreement to merge.

Both the Federal Energy Regulatory Commission and the California Public Utilities Commission investigated whether the proposed merger of Edison and San Diego Gas and Electric would facilitate increased evasion of rate regulation and foreclosure of competition. On behalf of the City of San Diego, Bruce Owen testified that the merger would do precisely that (Owen 1989, section III, and 1990). The Federal Energy Regulatory Commission's administrative law judge, the California commission's administrative law judges, and the California commission itself all concluded that the proposed merger raised serious competitive problems because it would facilitate evasion of regulation as contemplated by Section 4.23 of the 1984 *Merger Guidelines*.[10]

In its decision rejecting the proposed merger, the California commission stated:

> By expanding the geographic scope and extent of potential self-dealing, the opportunities of Mission Energy and the scope of SCEcorp's unregulated activities, the merger may increase the demands on Commission resources now devoted to affiliate issues. Indeed, if Edison's past violations of the regulatory compacts set forth in our holding company decision are any indication of what will transpire in the future, it will be increasingly difficult to ensure that

inappropriate costs are not passed on to ratepayers. (California Public Utilities Commission 1991b, 79)

The decision went on to say that the existing reasonableness review (referred to as ECAC) did not sufficiently protect ratepayers.

The ECAC's rapid schedule, its resource intensive nature, its reliance on the good faith provision of utility/holding company information, and its focus on fuel and purchase power costs (in this instance the QF purchase power contracts) to the exclusion of other significant affiliate-related issues (such as transmission access) make the ECAC an unsuitable forum to protect against the adverse impacts identified in this decision. The Commission's pre-merger experience is that the ECAC has not been an ideal forum for adjudicating contested affiliate issues. This casts serious doubt on the ability of this mechanism to resolve the greater problems associated with the expanded self-dealing opportunities facilitated by the merger. (California Public Utilities Commission 1991b, 92)

Following the California commission's decision rejecting the proposed merger, the merger was abandoned.

COMPETITIVE BIDDING

Edison claimed that the California commission's system for entering into long-term standard offer contracts on a competitive bid basis would prevent abuse when Edison contracted with an affiliated qualifying facility (Southern California Edison 1989, 6-2 to 6-11). A competitive bidding system, however, would be unlikely to prevent the types of abuses discussed in this chapter.[11] This may explain why in eleven of twenty states in which utilities have solicited competitive bids for generation, investor-owned utilities, their affiliates, or both are excluded from bidding (*Electric Utility Week*, Mar. 8, 1993, 17).

We have identified four problems with reliance on competitive bidding to prevent abuses of self-dealing. First, a purchasing utility could probably favor affiliated qualifying facilities in awarding contracts even when competitive bidding is used. Notwithstanding various safeguards, there is likely to be scope for the utility to favor affiliated bidders through the specification of the terms of the contracts put up for bid, through provision of inside information, and through subjective aspects of bid evaluation. The latter include the way the utility assigns scores to each project for each of the criteria used in bid evaluation and the weights it uses to combine the various criteria to produce one ranking. In rejecting the proposed California utility merger, the California commission stated that "competitive bidding in its various forms is heavily dependent upon utility input, and does not remove all subjectivity from Edison's contract award decisions, thus raising the question of whether it is a sufficiently neutral vehicle to protect ratepayers, notwithstanding applicants' claims" (California Public Utilities Commission 1991b, 93). A Maryland Public Service Commission hearing examiner recommended against approval of Delmarva Power and Light's participation in a cogeneration project because it appeared that the utility's bidding proposal, including its scoring system, was weighted toward the project (*Electric Utility Week,* June 3, 1991, 3).

Second, competitive bidding would not prevent the purchasing utility from engaging in various abuses in the course of contract administration and amendment after contracts are awarded (Owen 1989; Ahern 1985; California Public Utilities Commission, Division of Ratepayer Advocates 1988). When deciding whether to exercise options in its contracts with affiliated qualifying facilities and how to respond to changes in circumstances that might involve contract renegotiation and amendment, a utility could base its decision not on its ratepayers' interests but on whether the action would raise or lower the profits of the affiliated qualifying facilities. Indeed, as the California commission stated, "The competitive bidding . . . protocol only addresses the initial contracting stage, not the contract administration stage, where abuses can raise the specter of substantial excess ratepayer costs" (California Public Utilities Commission 1991b, 92–93).

Third, the California commission has suggested that utilities could purchase additional performance features from qualifying facilities after contracts were awarded. An example of such a performance feature would be control by the purchasing utility over level of generation by the qualifying facility. Even if some of the benefits would go to nonaffiliated qualifying facilities, a utility such as Edison that has contracts with affiliated qualifying facilities might benefit by overpaying for such added features (California Public Utilities Commission 1986b, 2, 74–75).

Fourth, the California commission allows postbid contract negotiation between utilities and affiliated qualifying facilities (California Public Utilities Commission 1986b, 71, 75, 100; 1987, 45–49). With such negotiation, discrimination in favor of affiliated qualifying facilities would be possible. Elimination of discretion in the procurement process (for example, through use of standard offers, and through procedures that do not allow postbid negotiation) is not a reasonable way to deal with this problem, because it would be likely to raise the cost of generation. Efficient long-term purchase contracts are inevitably complex, and this complexity suggests that the parties involved should be free to negotiate the terms of the contract. Joskow suggests that "there are clear practical problems with rigid self-scoring competitive bidding systems, which may already be emerging in Massachusetts. Boston Edison, the company farthest along with a highly structured bidding system that has little room for bilateral negotiations, has run into problems with several of the projects selected through bidding" (Joskow 1989b, 183; see also Joskow 1991, 82; 1992, 30–31). Thus there is a conflict between freedom to negotiate efficient contracts and utility ownership of qualifying facilities from which they purchase power.

UTAH AND NORTHEAST UTILITIES

Power purchases from affiliated qualifying facilities are not the only context in which vertical mergers in the electric utility industry have raised concerns about evasion of rate regulation and foreclosure of competition. In his prefiled testimony in *Utah Power and Light*, Hughes discussed how a vertical merger between one firm with generating interests, Pacific Power and Light, and another firm with transmission interests, Utah Power and Light, could facilitate the evasion of Federal Energy Regulatory Commission regulation of transmission rates.

> A regulated, vertically integrated firm having market power in one aspect of its business may seek to avoid regulatory scrutiny by packaging its product in one bundle so as to hide the non-competitive price inherent in one of the components of that bundle. . . .
>
> As a practical matter, regulatory scrutiny of bulk electricity transactions of a vertically integrated company is unable to identify the profits on transmission that are implicit in the price of the bulk electricity. . . .
>
> [T]here are strong indications that the merged company will avoid FERC scrutiny by bundling UP&L transmission with PP&L-generated bulk electricity. (Hughes 1988, 7, 24, 26–27, 54)[12]

The Federal Energy Regulatory Commission found that the merged firm that controlled Utah Power and Light's transmission facilities and Pacific Power and Light's generation facilities could deny transmission service to the cheapest power that would be available to supply the Southwest. Instead, the merged firm could supply the Southwest from Pacific Power and Light's more expensive generating facilities. Such foreclosure of competition would facilitate evasion of regulation of transmission rates and thus the exercise of increased market power in transmission (Federal Energy Regulatory Commission 1988, slip op. at 27–37).

The Federal Energy Regulatory Commission found a similar problem in connection with the merger of Northeast Utilities (NU) and

Public Service of New Hampshire. Absent mandatory transmission access conditions, the proposed merger would have increased the merged firm's market power in transmission in certain corridors. Independent of this market power in transmission services, the proposed merger would have increased the merged firm's market power in short-term bulk power (Federal Energy Regulatory Commission 1991a, slip op. at 38). In addition to these horizontal competitive problems, the Commission found a vertical competitive problem. The increased market power in transmission that would result from the proposed merger would enable the merged firm to deny transmission access and force customers in New England to purchase bulk power from the merged firm at prices above those at which power was available from alternative sources.

> NU's control of key transmission corridors and facilities would allow it to control bulk power trade. Its substantial inventory of excess generating capacity would give NU the incentives to block the sale of competing sources of short-term bulk power services. To the extent that NU's resources are not the lowest cost resources that would otherwise be available, the results would be higher electricity prices for consumers in many parts of New England. (Federal Energy Regulatory Commission 1991a, slip op. at 43; see also 22–23, 26, 38, 40)

CONCLUSION

When two unregulated firms propose to merge, the antitrust analysis of the transaction normally focuses on the markets in which the two firms compete, that is, horizontal issues. However, when at least one of the firms proposing to merge is subject to regulation, it is important to consider whether the merger would be likely to permit evasion of regulations that constrain the exercise of market power. The most common area of concern involves mergers between monopoly firms subject to cost-based regulation and firms in competitive markets. Particular concerns arise when the firms are also vertically related, that is, when one of the merging firms sells goods or services to the other.

Two antitrust problems can arise in the case of a vertical merger between a monopoly subject to cost-based regulation, such as an electric distribution company, and a competitive supplier of inputs, such as a generating company. First is evasion of regulation. Through improper allocation of costs to the regulated activity, or through inflation of prices paid by the regulated enterprise to the competitive one, the merged company may be able to obtain regulatory approval to increase prices charged by the regulated monopoly. Second is foreclosure of competition. To evade regulation the monopolist has an incentive to purchase inputs from its competitive affiliate even if the latter is not the supplier with the lowest costs of production.

Numerous legal and regulatory investigations involving telephone services and electric utilities have found that evasion of regulation and foreclosure of competition by regulated monopolies are real problems. The Department of Justice's concern regarding this problem is stated in the 1984 *Merger Guidelines*, and both the Antitrust Division and the Federal Trade Commission have used the evasion-of-regulation theory in litigation.

NOTES

1. This chapter is concerned primarily with antitrust problems that arise when the competitive activity is unregulated. See Braeutigam and Panzar 1989 for an analysis in which a firm with a regulated monopoly has an incentive to overexpand in a competitive activity in which it is also subject to regulation, even when regulators allow a rate of return that does not exceed the actual cost of capital. In Averch and Johnson (1962), a similar incentive exists only if the rate of return allowed by regulation exceeds the true cost of capital. Antitrust concerns may also arise if a firm has regulated monopolies in two activities that are subject to different rate regulation formulas.

2. Even in the case of rate-based generating capacity, there are incentives to foreclose competition in generation. At the margin, investor-owned utilities can pass some of the profits from wholesale transactions to shareholders rather than to retail customers (Acton and Besen 1985, vi, 35–37; *Electric Utility Week*, Nov. 16, 1992, 11). Investor-owned utilities may fear that greater excess rate-based generating capacity will lead to tougher regulatory treatment by state commissions (Federal Energy Regulatory Commission 1989a, 72).

3. Owen (1989, Section III.A) discusses situations in which similar problems arise when the upstream activity is regulated but competitive. In some situations, the actions of the vertically integrated firm might entirely foreclose competitors from the otherwise competitive upstream activity.

4. The 1992 *Horizontal Merger Guidelines* did not replace the sections of the 1984 *Merger Guidelines* dealing with competitive problems from vertical mergers. See U.S. Department of Justice and Federal Trade Commission, "Statement Accompanying Release of Revised Merger Guidelines," Washington, D. C., April 2, 1992.

5. 109 FTC 1986, complaint at ¶¶20, 21, 24; see also Calvani 1987, 6. The evasion-of-regulation theory was used by the U.S. Department of Justice (1982, 52–55) in analyzing coal purchases, and a version of the theory was used by the Department in analyzing Gulf Coast deepwater ports and the Alaska Natural Gas Transportation System (Owen 1989, 51–53).

6. U.S. Department of Justice, *Response to Public Comments on the Proposed Modification of Final Judgment*, 47 Fed. Reg. 23,320-336, (1982), quoted in *United States v. Western Elec. Co.*, 673 F. Supp. 525, 568 (D.D.C. 1987). See also quotes

from William Baxter in Collins and Loftis 1988, 83–87, and Baxter 1983, 243–47.

7. See also U.S. Department of Justice, "Report and Recommendations of the United States Concerning the Line of Business Restrictions Imposed on the Bell Operating Companies by the Modification of Final Judgment," Feb. 2, 1987, which suggests (at 49) that the Modification of Final Judgment itself "mirrors this more narrow antitrust focus by establishing the probable effect on competition in the markets a BOC seeks to enter—rather than the effect on customers in the regulated market—as the legal standard for removal of the line of business restrictions." Foreclosure could affect third-party customers by raising the costs of competitors when there are economies of scale, with the result that prices could increase in other markets.

8. For other examples, see Cross 1992.

9. The California commission issued standard offer contracts that utilities could use in purchasing power. These specified minimum requirements for performance and for various characteristics such as dispatchability, along with penalties and bonuses. Utilities were also allowed to use other—or non-standard—contracts. However, the standard offer contract formed the benchmark against which the reasonableness of all utility actions relating to qualifying facility transactions was measured (California Public Utilities Commission 1990, 119–21).

10. 53 FERC ¶63,014, slip op. at 39-51, Nov. 27, 1990. The Justice Department concluded that the proposed vertical merger of San Diego Gas and Electric and Mission Energy raised competitive concerns relating to evasion of rate regulation by the merged company. However, the Department's concerns evidently arose because Edison had contested the California Public Utilities Commission's jurisdiction to regulate its transactions with Mission Energy. The Department was willing to accept as a remedy an agreement on the part of the merged firm that affiliate transactions would be subject to prior approval by the California Public Utilities Commission, or, with the commission's approval, would be subject to competitive bidding. In the Department's view, this would allow the commission to determine whether it was able to regulate Edison/Mission Energy transactions and to impose any restrictions or prohibitions it wished on such transactions (U.S. Department of Justice 1990, 3–6).

11. Systems for bidding by qualifying facilities to supply capacity are discussed in Plummer and Troppmann 1990.

12 . Hughes notes that the Federal Energy Regulatory Commission had not prevented Utah Power and Light's unilateral exercise of market power in transmission prior to the merger.

7

Utility Mergers and Retail Competition

This chapter discusses the nature, extent, and benefits of retail competition in the electric power industry. Two types of retail competition—competition among electric distribution companies and competition between natural gas and electric distribution companies—are important for merger analysis. A third source of retail competition faced by electric distribution companies is self-generation by large customers.

Effects on retail competition between electric distribution companies may be an issue in connection with three types of mergers: a "downstream" horizontal merger between electric distribution systems; a vertical merger between a transmission or generation company and an electric distribution system; and an "upstream" horizontal merger between a vertically integrated investor-owned utility and another transmission or generating system. We will consider each of these types of mergers.

This chapter will also analyze how retail competition is affected by a merger between gas and electric distribution companies. It will not specifically address other types of gas-electric mergers that could affect retail competition, such as an "upstream" merger between a vertically integrated electric utility and a gas pipeline.

RETAIL ELECTRIC COMPETITION

Analyses of retail competition between electric companies typically deal with up to five forms of competition: franchise competition, head-to-head competition, fringe area competition, industrial location competition, and yardstick competition. We will discuss each of these in turn.

Franchise Competition

The term "franchise competition" is used in connection with three types of threats that could constrain the retail pricing of an electric distribution system, whether it is part of an investor-owned utility or is municipally owned.

Replacement of One Investor-Owned Utility by Another. First, there is the threat that a municipality might award its electric distribution franchise to a different investor-owned utility. Such a threat can be weakened or removed by a merger between the utility with the distribution franchise and another utility that would be a potential alternative operator of the distribution system.

Municipalization. Second, there is the threat that a municipality might take over a distribution system owned by an investor-owned utility. A merger that involved nothing more than two distribution systems would not directly affect such a threat, but the merger of two vertically integrated utilities, one of which serves the municipality, could make it more difficult for a municipally owned distribution company to obtain bulk power on competitive terms (Taylor 1989, 47–48). This could weaken the constraint imposed by the threat of municipalization.

Replacement of Municipal Ownership by Investor Ownership. Finally, where there is a municipally owned distribution system there is the threat that voters might award the distribution franchise to an investor-owned utility. It has been suggested that a merger between two utilities, each of which was a potential alternative operator of the distribution system, might weaken such a constraint.

There is controversy over the competitive significance of each of these three types of threats. Joskow (1985, 206) states that although franchise competition between investor-owned utilities could yield significant efficiency gains, opportunities for replacement of one investor-owned utility by another "are for all intents and purposes nonexistent" because of state laws and regulatory procedures. The overall assessment of Pace and Landon is consistent with Joskow's. They observe, however, that "when the first utility's franchise expired, presumably there would be no obstacle to enfranchising another private utility" (Pace and Landon 1982, 58).

Joskow (1985, 206) states that although municipal takeovers are possible in some states, they are "almost impossible" in others. Pace and Landon (1982, 57) argue that because franchise changes involve substantial transactions costs, they occur rarely and only when there is a clear long-term benefit. They report that between 1960 and 1978 twenty-seven franchises were transferred from private companies to municipal systems. They also report that these takeovers were motivated by factors such as preferential access to subsidized bulk power for municipal systems and not by concern about the exercise of market power.

Of course, one cannot infer from a low number of actual takeovers that the threat of such a takeover does not constrain retail pricing. The threat may be effective and hence seldom carried out. In their decision on the merger between Southern California Edison and San Diego Gas and Electric, the California administrative law judges found "the evidence indicates that the threat of potential municipalization is an important spur to utilities such as SDG&E and Edison to operate efficiently and reduce costs" (California Public Utilities Commission 1991a, 885).

Albuquerque provides an example of how the first two types of threats may be used to hold down electricity prices. Public Service of New Mexico's distribution franchise for Albuquerque expired in 1992. Albuquerque considered municipalizing its system and obtaining power from El Paso Electric or Plains Electric Generation and Transmission Cooperative, which had excess generating capacity and transmission lines within ten miles of the city, or alternatively awarding the franchise to Southwestern Public Service, which supplied power 150 miles away. However, Public Service of New Mexico refused to provide "retail wheeling" that would allow other utilities to supply power to Albuquerque. A bill was introduced in the state legislature to require retail wheeling in New Mexico. Public Service of New Mexico and the city of Albuquerque reached a tentative agreement for a franchise renewal that included a pledge not to increase rates for two years (*Electric Utility Week*, Dec. 14, 1992, 4–5; Jan. 18, 1993, 1, 8).

The 1992 Energy Policy Act has increased the potential scope of municipalization. It provides municipal distribution systems with greater transmission access to sources of bulk power other than adjacent or surrounding investor-owned utilities.

While maintaining that on balance franchise competition is almost nonexistent, Joskow (1985, 207) indicates that the third type of threat—that voters will transfer distribution franchises from municipally owned systems to investor-owned utilities—does constrain unregulated municipally owned systems to operate more efficiently and limits their ability to charge monopoly prices for electricity to supplement local tax revenues. According to Pace and Landon (1982, 58), ninety-two municipal systems were acquired by investor-owned utilities between 1960 and 1978. Most of the acquisitions were small, and all but twenty-five occurred before 1970. (Pace and Landon do not interpret these data as evidence that franchise competition promotes efficiency.)

Some might argue that because franchise decisions are based on long-term expectations, they would not be affected by, and hence could not prevent, short-term exercise of market power. Even if this were true, however, a constraint on the long-term exercise of market power would be important. Because industrial location decisions are also long-term, a similar argument may apply to industrial location competition, which is discussed later. If the relevant product market is long-term contracts to supply power to new or relocating plants, then industrial location competition could constrain even short-term exercise of market power.

Head-to-Head Competition

Even in the absence of laws and regulations limiting head-to-head competition, one might not expect to find competing electric distribution systems serving the same territory. Distribution of electricity in a given area has natural monopoly characteristics. At least for small customers that consume electricity at low voltages and for distribution companies that minimize costs, one would expect the cost of distribution to be lower when there is a single distribution company. For example, there are economies from avoiding duplication of distribution lines and equipment. These economies, however, may not be so substantial that they prevent actual and potential competition, particularly if firms that do not face competition fail to minimize their costs. In 1966 there was direct competition between two electric distribution companies in forty-nine cities with populations of 2,500 or more (Primeaux 1975a, n. 6). This is roughly the same degree of head-to-head competition as occurs in the cable television industry.

Head-to-head competition is almost always precluded by state and local laws that, at least de facto, provide distribution systems with territorial exclusivity (Joskow 1985, 178–79, 206; see also n. 86). One rationale for such laws is to prevent a loss of economies of scale. Another is that head-to-head competition would conflict with other government objectives for retail pricing. A goal of universal service at uniform prices is likely to involve cross-subsidies among customers, with some paying more than their costs. Competition would eliminate the profits earned from such customers and therefore the willingness of the distribution companies to supply customers that pay less than their costs. Such subsidies, an important issue in the telephone industry, are less important in electric power distribution.

In the future there would be significant head-to-head competition for large industrial buyers of high-voltage power if utilities were allowed to sell power to these customers located in other companies' service areas (Owen 1989, 179–81, Joskow 1986, IV-9 to IV-11; but see Joskow 1990, 139–42). This would require an end of territorial exclusivity and mandatory wheeling for such customers. It would be facilitated by deregulation of rates where competition is adequate.

Fringe Area Competition

Fringe area competition between distribution companies refers to two kinds of competition: competition to serve new developments in nonfranchised areas located between existing franchise areas, and competition to serve previously undeveloped areas near the boundary between franchise areas. According to Joskow (1985, 212–13), the first type of competition is permitted in a few states, but it is fairly rare and is not generally important in the electric power industry. Competition of the second type would require changes by state regulatory commissions in boundaries between franchise areas. Such boundary changes may be difficult to obtain in response to differences in prices or service levels.[1]

Fringe area competition directly involves only a portion of a distribution entity's retail loads. Regulation limits the extent to which utilities can offer special prices to particular areas where the elasticity of demand is relatively high because there is an alternative supplier. However, because they are not subject to rate regulation, municipally owned distribution systems may be able to practice price discrimination. Furthermore, uniform price regulation does not imply that fringe area competition has no effect. Insofar as pressures to reduce costs are concerned, such regulation might even magnify the cost reduction. Insofar as the exercise of market power is concerned, this regulation would cause fringe area competition to have a smaller effect per kilowatt-hour over a larger area than would be the case if price discrimination were permitted. Similar points apply to industrial location competition.

Industrial Location Competition

To attract new customers, to prevent the relocation of existing customers, and to induce large—usually industrial—customers to locate in their territories, distribution systems discount retail prices. The scope for this "industrial location competition" depends, among other things, on the electricity-intensity of customers' production processes. For most production processes, purchased electricity accounts for a small percentage of the total value of shipments. As a result, significant competition related to electricity is unlikely for most customers. There are, however, exceptions. In the mid-1980s, the cost of electricity as a percentage of shipment value was 27 percent for primary aluminum, 25 percent for industrial gases, 18 percent for alkalis and chlorine, 14 percent for electro-metallurgical products, and 11 percent for hydraulic cement (U.S. Department of Commerce 1988, tables 2, 4). Our experience indicates that firms producing primary aluminum and industrial gases shop for electric power in choosing locations for new plants.

For a merger significantly to reduce industrial location competition, the existing competition must involve relatively few utilities. A problem is unlikely to arise if an industrial customer can choose among the service territories of many utilities. But when the potential locations for a plant are constrained by transportation costs for inputs or outputs, as they are for industrial gases, there could be few competing utilities.

Yardstick Competition

Comparisons of the performances of different utilities may affect regulatory pressures to reduce costs and prices. This is known as yardstick competition. Comparisons of utilities might influence regulatory decisions concerning the costs that can be passed through to retail prices. Yardstick competition also could involve comparisons by customers or by utilities themselves that lead to efforts to reduce costs and prices. The existence of independent comparable utilities thus could make regulation more efficient, exert downward pressure on costs, and constrain the exercise of market power. Shleifer (1985) discusses the merits of rewarding similar regulated firms on the basis of relative performance, a system that he concludes can outperform cost-of-service regulation. Shleifer notes that cost comparisons across similar firms are not new to regulatory practice, and he cites Department of Defense use of dual-sourcing in hopes of saving enough from cost control to offset forgone economies of scale.

To implement this type of competition correctly, utilities should be comparable, or there should be adequate controls for important differences. Critics argue that observations for a large number of utilities are needed to control adequately for important differences. Therefore, the critics argue, whether or not any particular utility is available for the comparison does not make much difference. However, in assessing the existence of competition, the issue is not whether yardstick comparisons are done correctly but whether government officials, utilities, and the public use comparisons between the specific utilities in question in pressing for cost reductions.

In testimony supporting the proposed merger of Southern California Edison and San Diego Gas and Electric, Joskow (1990, 145) stated "inter-utility comparisons . . . do not play a significant role in determining retail rates." However, Joskow and Schmalensee (1983, 21–22) state that "[b]oth the FERC and some state regulatory agencies . . . have recently begun to use comparative data to evaluate utility efficiency." Joskow and Schmalensee also state that the effect of yardstick competition on costs and prices is "uncertain." In testimony on behalf of the merging parties, Pace (1989, 101) acknowledged that San Diego Gas and Electric "apparently has made a number of 'yardstick' rate comparisons in the past as a means of stimulating internal cost-control and rate redesign efforts and responding to customer pressure. . . . The merger may eliminate one source of downward pressure on SDG&E's costs and rates—comparisons with its neighbor."

In the state review of the same merger, the California commission's administrative law judges found significant yardstick competition between Southern California Edison and San Diego Gas and Electric, as well as between Edison and certain municipally owned systems. The judges stated that "SDG&E has used cost comparisons with Edison in its successful efforts to reduce retail rates." Moreover, the judges concluded that comparisons between the two California utilities were useful to state regulators in measuring the reasonableness of purchased power costs because "SDG&E's system is the system most comparable to Edison's due to its contiguous Southern California location, and its access to the same bulk power markets." The judges concluded that "regulators may also use SDG&E as a source of information to test whether Edison's generation facilities are efficient and competitively priced" (California Public Utilities Commission 1991a, 880, 882–83).

The California Public Utilities Commission (1991a, 110–11) found that the proposed merger would result in loss of "across-the-fence rivalry." This rivalry had "a perhaps indirect, but indisputably important, influence on quality of utility service" as well as cost of service to ratepayers. The "across-the-fence rivalry" described by the California commission includes yardstick competition, although the commission rejected the latter term.

Extent of Competition

The extent of the various types of competition among electric distribution companies is subject to much debate. In the context of litigation, those working on behalf of investor-owned utilities find the competition minimal. Regulators and the courts, however, have found retail competition to be significant.

Pace and Landon (1982) suggest that head-to-head, industrial location, and yardstick competition are nonexistent or are of no importance. They acknowledge that a very limited amount of franchise and fringe area competition may exist, but they conclude that it does not contribute to efficiency.

Joskow (1985, 185) argues that retail competition is almost nonexistent.[2] He suggests that litigation alleging reductions in retail competition is largely a product of efforts by municipally owned systems to obtain concessions from investor-owned utilities. Concerns about nonexistent retail competition have deterred mergers and acquisitions, he argues. In particular, acquisitions of small, municipally owned systems and utilities that could result in lower costs have been deterred (Joskow 1985, 174–75, 225).

By contrast, in the context of specific mergers and other antitrust cases, the federal courts, federal regulators, state regulators, and various expert witnesses have found retail competition to be significant.[3]

Effects of Competition on Costs

Independent entities that distribute electricity may constrain each other's costs, retail pricing, and quality of service. Entities with low costs or prices may replace those with higher costs and prices.

In the mid-1960s, municipally owned distribution companies that faced head-to-head competition from privately owned companies had significantly lower costs than did similar municipally owned distribution companies that did not face such competition (Primeaux 1975a). Primeaux's sample consisted of forty-eight distribution companies, twenty-three of which faced head-to-head competition. This finding applies only up to an annual output of 222 gigawatt-hours, which is very small.

Joskow and Schmalensee (1983, 61–62) point out that Primeaux's work does not necessarily imply that two competing systems serving a given area have lower average costs than a single system. Primeaux's estimate of the effect of competition held the customer density of a system constant, but the density would not remain constant if overlapping systems were merged. Primeaux did not examine whether the net effect (allowing for the effect of competition on density) is that two competing systems have lower costs. Pace and Landon (1982, n. 150), Hamilton and Hamilton (1983), and Joskow and Schmalensee also criticize other aspects of Primeaux's work. Nonetheless, Primeaux's study does provide support for the hypothesis that retail competition causes distribution companies to take steps that reduce their costs.

Effects of Competition on Retail Prices

Because state regulation constrains retail electricity prices, any retail competition that might exist would have no significant effect on prices, according to Joskow (1985). We do not believe that rate regulation eliminates the role of competition. For example, retail competition could constrain the ability of an investor-owned utility with a financial interest in an unregulated generating facility to evade regulation of retail rates by purchasing power at inflated prices from its affiliate.[4] Even if cost-based regulation were effectively to constrain average prices, it could allow rate structures that involve substantial departures from efficient prices (for example, retail prices of electricity that do not depend on season and time of day). Competition could bring about more efficient rate structures.

Efficiency of Competition

Pace and Landon (1982) and Joskow (1985, 217–18) conclude that retail competition plays essentially no role in promoting economic efficiency in the electric power industry.[5] Part of their argument is that such competition is almost nonexistent. Furthermore, Pace and Landon (1982, 56–57) and Joskow (1985, 213) argue that competition may have perverse effects where regulated retail prices depart from marginal costs. Competition might lead to replacement of a low cost supplier by a higher cost one.

Mergers Affecting Competition

As noted earlier, three types of mergers can affect retail competition between electric distribution companies: a "downstream" horizontal merger between distribution systems; a vertical merger between a transmission and/or generation firm and a distribution system; and an "upstream" horizontal merger between a vertically integrated investor-owned utility and another transmission and/or generating system.

The way in which a "downstream" horizontal merger can affect competition is relatively straightforward and will be discussed later. We will now consider the second and third types of mergers.

A vertical merger and an "upstream" horizontal merger have an important feature in common. Each might lead to a situation where a firm that is vertically integrated into distribution would have market power upstream in transmission and/or generation. Retail competition may be greater when competing distribution systems have access to bulk power on equal terms. Retail competition may be less when one of the distribution companies is owned by a firm that has a monopoly over bulk power supplies to the other. In their proposed decision on the California utility merger, which for present purposes was a merger of the third type, the California administrative law judges found that the adverse effect of the merger on wholesale bulk power and transmission services would reduce retail competition between Edison and certain municipally owned systems that were partial-requirements customers (California Public Utilities Commission, 1991a, 885). If there is retail competition, this finding is consistent with suggestions made by Joskow (1985, 211):

> If this kind of industrial location competition actually is of some significance under current institutional arrangements (a dubious proposition), it is most likely to be an important spur to lower costs and a constraint on monopoly power when each of the competing distributors has its own sources of generating and transmission capacity either because both are integrated or because one or both have access to alternative wholesale suppliers who, in turn, are in competition with each other and wholesale prices are determined as a consequence of that competition.

To exploit its monopoly power more fully, the upstream monopolist may have an incentive to take over the independent distribution company. It might do so to evade regulation, if there are unregulated affiliated generating facilities, or to increase the rate base, if the allowed rate of return exceeds the cost of capital. A takeover also would eliminate a problem of "double marginalization" and might permit price discrimination. (Double marginalization occurs when a wholesale supplier with market power sells to an independent retailer with seller market power. Each charges a price above its marginal costs. As a result, the two firms taken together do not maximize their combined profits, and consumers are worse off than with a single, vertically integrated monopolist.)

While elimination of double marginalization is efficient, elimination of the independent distribution company could reduce efficiency. Retail prices might increase because of the exercise of market power, and as a result of regulation the integrated monopolist might have higher distribution costs than a municipal distribution system.

Efforts by the upstream monopolist to take over the independent distribution company could involve pricing and other policies that would limit the ability of the latter to compete at retail. In this case, to the extent that they contributed to the formation of an upstream monopolist that is integrated into distribution, mergers of the second and third types might reduce retail competition even though they do not directly increase concentration in retail markets.

Market Definition

In mergers between electric distribution companies, an obvious starting point for product market definition is retail electricity. However, a broader product market, such as retail energy (electricity, natural gas, propane, and fuel oil) might be relevant. Also relevant might be a narrower product market, such as retail electricity for residential customers. Another possibility is that the product market would be broader in some respects and narrower in others (for example, retail energy for large industrial customers).

At least in the case of electricity and gas, tariff structures and regulations and state barriers to resale leave room for price discrimination between different categories of customers, such as residential versus industrial. Moreover, competitive conditions are likely to vary among customer categories. For example, the cross-price elasticity of demand between electricity and natural gas is likely to differ for residential and industrial users (Joskow and Baughman 1976; Beierlein et al. 1981; Blattenberger et al. 1983; Cohen et al. 1985). (The cross-price elasticity of demand for electricity with respect to the price of gas measures the responsiveness of the quantity of electricity purchased to changes in the price of gas. It is the percentage change in the quantity of electricity purchased that would result from a one percent increase in the price of gas.) Furthermore, the cross-price elasticity of demand between electricity and gas is likely to differ across residential uses (for example, cooking versus space heating). Thus different customer categories, and to some extent different uses, could be in different product markets.

Although natural gas, propane, or fuel oil are likely to impose a constraint on retail electricity prices for some users and uses, in general they would not make "small but significant and nontransitory" retail price increases for electricity unprofitable. This is confirmed by estimates of the price elasticity of demand for electricity, which indicate that for broad categories of customers at prevailing price levels a firm with a monopoly over retail electricity would earn higher profits if it raised prices. (Joskow and Baughman 1976; Beierlein et al. 1981; Blattenberger et al. 1983; Cohen et al. 1985). At least for important categories of users and uses, retail electricity is a relevant product market for analysis of competitive effects of mergers between electric distribution companies.

If the various types of retail competition were considered separately, the geographic market for retail electricity would be likely to vary among the types of retail competition. (For example, industrial location competition might take place over a different area than yardstick competition.) However, since uniform tariff schedules apply to a given category of customers throughout a distribution company's service area, it is probably most sensible to view each affected company's service area as a relevant geographic market and to evaluate the effect of the merger on competition in that area. There is, however, a complication: to the extent that the tariff schedules of two merging companies are to be harmonized, the merger would not only eliminate competition between the merging companies but also combine the remaining competition in the two areas. In principle, the latter effect of a merger could produce competitive benefits.

It is important not to lose sight of the objective of market definition. One should begin with an idea of the form that a possible exercise of market power might take (for example, a general price increase or a more focused price increase involving price discrimination). The purpose of market definition is to identify the sources of competition (customer alternatives) that would prevent a "small but significant and nontransitory" exercise of market power prior to the merger. This information is then used to determine whether the merger would reduce competition significantly.

In analyzing the effect of a merger of two electric distribution companies on retail competition, sophisticated concentration indexes such as the Herfindahl-Hirschman Index are not likely to be illuminating in most cases.

COMPETITION BETWEEN GAS AND ELECTRICITY

Natural gas and electricity are distributed by separate companies in some areas of the United States and by a single combination utility in other areas. In 1967, 43 percent of the largest privately owned electric utilities in the United States also distributed natural gas (Owen 1970, 716). The U.S. Senate introduced a bill in 1970 to prohibit combination gas and electric utilities, and in the early 1970s a number of studies compared the performance of combination and separate utilities. The bill did not become law, and mergers between gas and electric distribution companies with overlapping service territories have been permitted. For example, in 1983 Kansas Power and Light acquired Gas Service Company, and in 1985 Public Service of New Mexico acquired the Gas Company of New Mexico.

Particularly for residential customers, gas and electricity compete in significant uses. The extent of this competition depends on the relative prices of gas and electricity and hence varies over time and among locations. Residential customers choose between electric and gas appliances (ranges, clothes dryers, water heaters, central air conditioners, space heaters) based on the relative prices of gas and electricity (Landon and Wilson 1972, 259–60). Studies of the demand for energy indicate that gas and electricity are substitutes (Joskow and Baughman 1976; Beierlein et al. 1981; Blattenberger et al. 1983; Cohen et al. 1985). Electric and gas distribution companies that are owned separately engage in greater sales promotion than do separate utility companies (Landon and Wilson 1972, 261; Landon 1972, 12).

Although retail electricity is a relevant product market for many users and uses, and therefore is appropriate for use in analyzing mergers between electric companies, there is a broader product market for retail energy including gas and electricity that is relevant for mergers between gas and electric companies. It may well be that the relevant retail energy markets should be defined for particular customer groups (for example, residential users) and that in some situations they would involve other energy sources such as propane or fuel oil.[6]

Because combination utilities face less competition, their costs, prices, and profits may be higher than are those of separate gas and electric utilities. Insofar as prices and profits are concerned, this hypothesis is based on the premise that regulation is ineffective in eliminating the exercise of market power.

Cost Studies

A number of studies have compared the costs of combination and separate distribution companies. The central issue is whether any cost reductions that result from competition outweigh any economies of scope that are achieved when a single company distributes both gas and electricity and consolidates activities, such as meter reading and billing.[7]

Sing (1987) analyzed 1981 data for 108 distribution companies. Compared with separate companies, combination companies had lower costs at some output levels and higher costs at others. At the average output levels for the combination companies in the sample, Sing found that combination companies had 7 to 8 percent higher costs. Depending on output levels as well as the magnitude of any merger-related costs, her results suggest that some mergers between gas and electric distribution companies would reduce costs while others would raise them. The model Sing uses assumes that distribution companies minimize costs. Thus higher costs for combination companies are interpreted as resulting from diseconomies of scope rather than from reduced efforts to control costs.

Based on 1979 data for 200 distribution companies, Mayo (1984) concluded that at relatively low output levels for gas and electricity, combination companies had slightly lower costs (less than 1 percent lower). At higher levels for either or both outputs, separate companies had lower costs (sometimes as much as 12 percent lower).

Stevenson studied 1972 data for seventy-nine electric utilities, of which twenty-five were combinations. He concluded that for utilities typical of those in his sample "a reduction in competitive pressure, stemming from the joint ownership of the electric utility and the local natural gas distributor would lead to higher costs for the generation of electricity on the magnitude of . . . 8.5%" (Stevenson 1982, 61). His study did not cover distribution costs for electricity or any costs for gas.

Price and Profit Studies

Several pricing and profit studies comparing combinations and separate electric utilities were carried out around 1970 based on data from the late 1960s. These studies do not permit definitive conclusions. For example, they do not permit rejection of the hypothesis that retail electricity prices charged by combination utilities were the same as those charged by separate electric companies (Pace 1972; Landon 1972, 1973, 1974). At the same time, the studies do not provide persuasive evidence that retail prices charged by combination utilities were not higher than they would have been for separate electric companies under the same conditions (Owen 1970, 1973; Collins 1973). Brandon (1971) found rates of return on the rate base for electric operations significantly higher for combination utilities than for straight electric utilities, but Landon (1972, 12–13) found no significant difference in return on equity.

Other Concerns

There has been concern that the combination of gas and electric distribution might lead to distortions in the relative use of gas and electricity as a result of the combination companies' efforts to promote whichever energy source was the most profitable under regulation. For example, if regulation allowed a return on equity higher than the cost of capital, this would provide an incentive for combinations to try to shift consumption toward the more capital-intensive energy.

Another concern may arise when a combination gas and electric utility supplies natural gas to an unaffiliated electrical generating plant with which it competes in the supply of electric power. Oxford Natural Gas, a gas distribution company, recently alleged that Cincinnati Gas and Electric has used its control of a natural gas pipeline to impede the development of a cogeneration project that would compete in supplying electricity (*Electric Utility Week*, Feb. 8, 1993, 1, 8–9).

CONCLUSION

Laws and regulations reduce the scope of retail competition in the electric power industry. Yet there is evidence of franchise, head-to-head, fringe area, industrial location, and yardstick competition in some circumstances. Where such competition exists, it is likely to contribute to lower costs and prices. An antitrust analysis of an electric utility merger must, therefore, evaluate effects on retail competition in the specific situation at issue.

Issues of retail competition also arise in connection with mergers between gas and electric distribution companies. A number of studies raise concerns about the economic performance of combination gas-electric utilities. None of the studies indicates that combination utilities generally achieve or pass on to customers significant economies as a result of combining gas and electric operations, although there is some evidence that under specific circumstances they might do so.[8]

NOTES

1. There is no fringe area competition of the second type in California, Pace argues. He believes a decision by the California commission concerning the

procedures for service territory boundary changes makes this clear (Pace 1989, 80–81; 1990, 120). Administrative law judges for the California Public Utilities Commission dispute Pace's interpretation of that decision, however (California Public Utilities Commission 1991a, 888, 890).

2. However, Joskow indicates that one type of franchise competition exists—the third type discussed in the text. Also, Joskow (1985) does not address yardstick competition. However, yardstick competition is discussed by Joskow and Schmalensee (1983, 21–22).

3. *Otter Tail Power Co. v. United States* 410 U.S. 366 (1973); *Federal Power Comm'n v. Conway Corp.* 426 U.S. 271 (1976); *City of Mishawaka, Ind. v. Indiana and Michigan Elec. Co.* 560 F. 2d 1314 (7th Cir. 1977). For a critique of these decisions, see Joskow 1985. For a more recent finding of retail competition (including industrial location competition), see *City of Anaheim v. Southern California Edison Co.* No. CV 78-810 MRP (C. D. Cal. 1990). On the findings of federal regulators, see *Connecticut Light and Power Co.*, Federal Energy Regulatory Commission Docket ERS78-517 (1979). For a more recent price squeeze case, see Federal Energy Regulatory Commission 1990b. Administrative law judges at the California Public Utilities Commission (1991a, 860–94) found there was yardstick, franchise, and fringe area competition. They gave the greatest weight to yardstick competition. For the view of one witness, see Taylor 1989, 31–52.

4. Costs inflated as a result of self-dealing would be included in the cost differentials in Primeaux's work, since the price of purchased power is not held constant. His "cost of purchased power" variable is actually the share of kilowatt-hours purchased.

5. Joskow and Schmalensee (1983, 20–23), who address yardstick competition as well as other types of retail competition, reach a somewhat different conclusion. They see no convincing empirical evidence that actual or potential retail competition leads to lower electricity prices.

6. However, in *Consolidated Gas Co. of Fla. v. City Gas Co. of Fla.* 665 F. Supp. 1493, 1517 (S. D. Fla. 1987), the court found it "clear that a five percent increase in the price of natural gas would not cause consumers to substitute LP [liquid petroleum-based] gas, as the price of the latter is already roughly 100% higher than the price of natural gas."

7. The study by Stevenson (1982) is limited to generating costs for electricity. The other studies cover all costs (at least for electricity) and have no way of controlling for possible differences in service levels (for example, tree removal and speed of response).

8. Landon (1972) suggests that combination utilities charge lower prices for industrial customers. However, this is based on one of four regressions for industrial prices, and the relevant standard error is so low compared with the others that one suspects an error. See also Owen 1973.

8
Remedies for Anticompetitive Mergers

When the Federal Trade Commission or the Antitrust Division of the U.S. Department of Justice determines that a merger may substantially lessen competition, there are three ways that merging parties may avoid having the antitrust agency attempt to block the transaction. Merging parties may sign consent agreements containing provisions that remove the antitrust agency's competitive concerns, convince the agency that cost savings from the merger would outweigh the adverse effects of increased concentration, or persuade the agency that one of the merging firms is failing and that the competition in question would be lost even if the merger were blocked. By far the most common of these are consent agreements.

The remedies accepted by the federal antitrust agencies in consent agreements differ from the transmission conditions commonly imposed by the Federal Energy Regulatory Commission in the case of electric utility mergers. As a rule, the federal antitrust agencies impose conditions designed to prevent a lessening of competition, and the agencies avoid conditions that require continuing regulation of the merged firm. By contrast, the Federal Energy Regulatory Commission allows mergers that will lessen competition, while imposing additional regulations on the merged firm in an effort to prevent it from exercising its increased market power. In fact, the Federal Energy Regulatory Commission's transmission conditions do not prevent the exercise of increased market power. Nonetheless, transmission conditions that are not sufficient to justify approval of a merger may be sufficient to warrant approval of market-based pricing.

In regulated industries, an unusual type of merger condition is available: restrictions on postmerger price increases. Such restrictions do little to alleviate competitive problems of electric utility mergers, however. The major competitive problems in electric utility mergers affect wholesale transactions, while the price restrictions apply to retail sales, and the restrictions on prices are temporary.

Chapter 9 analyzes the claims made by merging parties regarding the cost savings that would result from electric utility mergers. The failing firm defense (U.S. Department of Justice and Federal Trade Commission 1992, Section 5) has not played a role in the merger of electric utilities and is not addressed in this book. While some merging electric utilities, such as Public Service of New Hampshire, have sought protection under the bankruptcy laws, no one has expected that their assets would leave the industry.

ANTITRUST CONSENT PROVISIONS

The federal antitrust agencies frequently allow merging firms to settle Section 7 antitrust complaints by agreeing to conditions that are intended to prevent the mergers from increasing market power. Coate et al. (1993) report that merging parties entered into consent agreements without litigating in 40 percent of the prospective horizontal merger cases in which the Federal Trade Commission issued complaints during fiscal years 1984 through 1991.

Most consent agreements in horizontal merger cases are based on one of three strategies to prevent a reduction in competition: divestiture of assets by the merging firms to a purchaser that will be an independent competitor; removal of entry barriers and assistance to bring about the entry of an independent competitor; and provisions to make a joint venture partner behave as an independent competitor. Technology licenses and various other provisions are a normal part of consent agreements designed to remove entry barriers, and they are used in connection with divestitures as well.

Divestitures

Most consent agreements require the merging parties to divest certain assets. A proposed merger may substantially lessen competition in only one or some of the markets in which the acquired firm is a seller. In that case, the antitrust agencies may require divestiture of the assets used by one of the merging firms to supply the adversely affected market(s) while allowing the remainder of the acquisition to proceed.

The objective of a divestiture is to enable the purchaser to assume the competitive role formerly played by one of the merging firms, typically the one with the lower market share. The antitrust agencies require divestiture of whatever is necessary to make the purchaser of the divested assets a viable competitor. A divestiture requirement may include not only tangible assets used in developing, producing, and selling goods and services but also intangible assets such as proprietary technology, formulas and other trade secrets, test and product performance data, regulatory filings, copyrights, brand names, customer information, and distribution contracts.

A successful divestiture prevents a lessening of competition. However, divestitures do not always proceed smoothly after consent agreements are signed. The antitrust agencies often face problems obtaining timely divestitures that would establish viable competitors. Recent examples are the Federal Trade Commission's experiences with its 1990 orders against Institut Merieux and Red Food Stores (*FTC:Watch*, Mar. 22, 1993, 9; May 24, 1993, 22–23; June 7, 1993, 11–12).

Facilitation of New Entry

The antitrust agencies will not challenge a merger if they conclude that entry is easy. From time to time, merging parties are able to avoid a divestiture, or avoid an antitrust suit to block a merger when a divestiture remedy is not possible, by accepting conditions designed to facilitate entry of a new competitor. Typically, the purpose of these conditions is to bring about actual entry, not merely to make entry easy. Based on a review of consent agreements negotiated during 1990–92 by the federal antitrust agencies, Aronson and Keyte (1992, 27) found that a Clayton Act "Section 7 case premised primarily on the reduction in the number of competitors can sometimes be resolved by lowering critical entry barriers and stimulating rapid and effective market entry, as long as the consent itself does not substantially increase the likelihood of collusion or require undue administrative supervision."

Whether it is possible to settle a complaint without a divestiture depends on why entry is not easy under the *Merger Guidelines* standards. For example, if it would take more than two years to set up a new production facility because of the time required for construction of a plant and acquisition of machinery, one would expect the antitrust agencies to require divestiture of a plant. On the other hand, if production facilities can be set up easily and the limit on entry is merely access to patent rights, a nonexclusive license may be sufficient to make entry easy.

On occasion, merging parties agree to extraordinary measures to bring about new entry. The Federal Trade Commission recently allowed an acquisition on condition that the acquiring firm participate for four to six years as a minority partner in a production joint venture to be established with a new entrant. This was an interim measure until the new entrant was ready to operate alone (*Alpha Acquisition Corp. et al.*, 5 Trade Reg. Rep. (CCH) ¶23,004 (1991), *RWE Aktiengesellschaft et al.*, 5 Trade Reg. Rep. (CCH) ¶23,143 (1992)).

Joint Venture Partners

One type of joint venture is a company that maximizes its profits and pays dividends to its two parent companies. Such a joint venture acts as a single competitor. However, it is possible to set up a "competitive rules" joint venture with production in a single facility but with capacity, output, pricing, and marketing decisions made independently by the two parents, which are not permitted to share competitively sensitive information. Such a joint venture provides two competitors. (Chapter 2 discusses the latter type of joint venture in evaluating the scope for competition in transmission.)

When a merger raises a competitive problem, antitrust agencies occasionally accept a consent agreement with provisions designed to ensure that joint venture partners will be independent competitors. In one such case, the Federal Trade Commission, after attempting to obtain a divestiture, accepted an agreement by the acquiring firm not to restrict production, sales, or pricing in the United States by various joint venture partners. The agency assumed that these partners would provide competition to replace the alleged rivalry between the merging firms (*Hoechst Celanese Corp. et al.*, 5 Trade Reg. Rep. (CCH) ¶23,044 (1991)).

In 1984–85 the Department of Justice would not permit Alcan to acquire Arco's newly completed rolling mill because of concern about competition in aluminum can body stock. It required that the acquisition be replaced by a joint venture in which Arco would remain an independent competitor. The consent agreement allowed Alcan to acquire a 40 percent interest; Arco retained a 60 percent interest. The rolling mill was to be operated as a "competitive rules" joint venture for a period of ten years, with capacity and fixed costs allocated on a 40–60 basis. Each owner was to be solely responsible for determining, pricing, and marketing its output, each was to be responsible for paying the variable costs of its output, and each had the right to expand capacity unilaterally (*United States v. Alcan Aluminium Ltd. et al.*, 1985-1 Trade Cas. (CCH) ¶66,427, ¶66,428 (W. D. Ky. 1985)).

Licensing and Supply Agreements

Consent agreements designed to remove entry barriers normally require the merged firm to license technology to new entrants. Divestiture agreements sometimes require the merged firm to enter into similar licenses to help establish the purchasers of the divested assets as viable competitors. These licenses cover things such as patent rights, trade secrets, trade names, and customer lists (*American Stair-Glide Corp. et al.*, 5 Trade Reg. Rep. (CCH) ¶22,931 (1991)).

Similarly, the antitrust agencies may require that acquiring firms enter into agreements to supply inputs or outputs for resale to purchasers of divested assets or to new entrants for a period that may be as long as five to ten years (*United States v. General Binding Corp. and VeloBind, Inc.*, 1992-1 Trade Cas. (CCH) ¶69,727 (D.D.C. 1992)). The purpose of these supply agreements is to facilitate the development of an independent competitor.

Of course, high royalty rates or prices that would raise the licensee's variable costs of production could permit the licensor to exercise market power. To make licensing and supply agreements work as intended, the antitrust agencies must obtain agreement regarding how royalties and prices will be determined. For example, the agencies have required royalty-free licenses and product prices equal to costs.

In 1984 the Department of Justice concluded that the merger of LTV and Republic Steel was likely to reduce competition for certain products. In addition to divestiture of two plants, it required that LTV enter into a ten-year contract to supply the purchaser of one divested plant with stainless steel hot bands for rerolling. Presumably, this would enable the buyer to compete effectively in cold rolled stainless steel sheet and strip. The contract specified a price equal to LTV's input costs plus a fixed markup (*United States v. LTV Corp.*, 1984-2 Trade Cas. (CCH) ¶66,133 (D.D.C. 1984)). The contract, said to have been a nightmare, was abandoned.

Ancillary Provisions

Consent orders sometimes include a variety of additional provisions that are designed to make it easier for the purchaser of divested assets or a new entrant to compete. Examples include:

- A requirement that the merging parties provide personnel, training, and technical assistance for a limited period to help the purchaser get started (*Rohm and Haas Co. et al.*, 5 Trade Reg. Rep. (CCH) ¶23,195 (1992)).

- Requirements that the parties to the merger or acquisition revise contracts. For example, they may be required to reassign franchises or abrogate noncompete clauses (*Atlantic Richfield Co. et al.*, 5 Trade Reg. Rep. (CCH) ¶22,878 (1990)).
- A requirement that the acquiring firm assist in securing regulatory approval for a lessee to sell its product in the United States (*Institut Merieux S.A.*, 5 Trade Reg. Rep. (CCH) ¶22,779 (1990)).
- A prohibition against divesting firms hiring employees from the divested operations for some period of time.
- A prohibition against the merged firm entering exclusive or long-term sales or distribution contracts and otherwise preventing its distributors from selling competing brands (*American Stair-Glide Corp. et al.*, 5 Trade Reg. Rep. (CCH) ¶22,931 (1991)).

Conditions for Nonhorizontal Mergers

The preceding discussion relates to consent agreements in horizontal merger cases. Consent agreements in cases involving nonhorizontal mergers and acquisitions are different. For example, in 1984 the Department of Justice concluded that the combination of GTE's regulated monopoly telephone operating companies and Southern Pacific's unregulated competitive Sprint long-distance telephone service raised competitive problems of the type discussed in Chapter 6: evasion of regulation and foreclosure of competition. Based on distinctions between the facts in this case and in *United States v. AT&T*, the Justice Department agreed to the transaction subject to a number of conditions. The consent decree contained provisions intended to prevent cross-subsidization of the competitive activity by the monopoly activity and provisions intended to prevent GTE operating companies from discriminating in favor of the Sprint long-distance service. The consent decree also imposed requirements and a schedule for implementation of equal access for all long-distance carriers (*United States v. GTE Corp.*, 1985-1 Trade Cas. (CCH) ¶66,354 (D.D.C. 1984)).

FERC'S MERGER CONDITIONS

There are two principal distinctions between the conditions imposed on mergers by the federal antitrust agencies and the conditions imposed by the Federal Energy Regulatory Commission. First, the antitrust agencies usually seek *structural* remedies. They use a variety of strategies—principally divestiture requirements and provisions to facilitate new entry—to avoid reductions in competition. Second, while there have been exceptions, including a number of prominent mergers in the early 1980s, the federal antitrust agencies normally avoid conditions that involve additional regulation of the merged firm. By contrast, in a number of recent electric utility mergers, the Federal Energy Regulatory Commission has allowed a reduction in competition and an increase in market power, and it has imposed additional government regulation in an effort to prevent the merged firm from exercising that increased market power.[1]

Merger conditions of the type accepted by the Commission are inferior to the protection of competition through denial of a merger or divestiture. While regulation of the terms of access to transmission facilities may increase efficiency where there is market power, the preservation of existing competition, and of the potential for greater competition in the future, is likely to be better for customers than the creation of additional, albeit regulated, market power. Regulatory conditions often create incentives to behave inefficiently, and they raise the cost of supplying electricity. Enforcement of and compliance with continuing regulation are costly for taxpayers, regulated firms, and ultimately ratepayers. One need look no further than the dockets of regulatory commissions and the courts. And regulatory approaches typically do not prevent exercise of market power.

There is only one potential justification for the Federal Energy Regulatory Commission to accept a merger that reduces competition and to rely on increased regulation to mitigate the resulting increase in market power: persuasive evidence that the merger would result in efficiencies that could not be achieved without the merger. To justify the merger, these efficiencies must more than outweigh the costs resulting from increased regulation and greater exercise of market power.

It is important to distinguish between the Federal Energy Regulatory Commission's imposition of conditions in orders permitting market-based pricing, on the one hand, and in orders approving mergers, on the other. Conditions designed to limit the exercise of market power sufficiently to justify market-based pricing are not necessarily sufficient to justify a merger.

When the Commission evaluates market power in a decision on market-based pricing, the appropriate comparison is between conventional regulation and market-based pricing (combined with transmission conditions if necessary). The issue is to weigh the expected costs from the potential exercise of market power against the substantial costs involved in conventional regulation.

Merger cases present a different issue. In evaluating mergers, one must weigh the expected costs from the potential exercise of market power against the often disputed cost savings that the utilities could achieve only by merging. Because cost savings from eliminating regulation are more certain and may be greater than merger-specific cost savings, it is consistent for the Federal Energy Regulatory Commission to disapprove mergers in markets in which it would approve market-based pricing. The Justice Department has stated that it is appropriate to allow market-based pricing in some markets in which a merger would be regarded as anticompetitive (U.S. Department of Justice 1984a, 28–29).

Moreover, if it makes a mistake in approving market-based rates, or if competitive conditions change, the Federal Energy Regulatory Commission can reimpose regulation. By contrast, if the Commission makes a mistake in approving a merger, cost-of-service regulation will not restore the *status quo ante,* and consumers will have to live with the consequences.

ADEQUACY OF TRANSMISSION CONDITIONS

The principal antitrust problems identified in recent electric utility mergers are reductions in competition in transmission and bulk power. The Federal Energy Regulatory Commission has relied on transmission conditions to prevent increased exercise of market power. The Commission has favored "open access" transmission tariffs, which enable third parties to use any of the merged firm's transmission facilities on specified terms. However, when the competitive problem is limited to certain areas, the transmission conditions may be limited to particular corridors. In the future the Commission might conceivably rely on the transmission requirements imposed on all utilities by the Energy Policy Act of 1992 as sufficient to mitigate competitive problems.

The present section addresses the adequacy of the Federal Energy Regulatory Commission's transmission conditions to prevent the exercise of market power. We use as an example the Commission's 1993 decision not to investigate the potential competitive effects of the proposed merger between Entergy and Gulf States Utilities on the grounds that the open access transmission tariff offered by the merging parties eliminated concern over market power in transmission and bulk power (Federal Energy Regulatory Commission 1993a).[2] The Commission's reasoning was flawed. If the Entergy/Gulf States Utilities merger would lead to an increase in market power, the proposed transmission conditions would not prevent the merged firm from exercising that power.

Under the open access transmission tariff proposed by the merging parties, many entities—utilities, generators of power, and local distribution systems—would be able to obtain wheeling using the combined

Entergy/Gulf States Utilities transmission facilities. The rate for *firm* transmission service would be based on the combined system's embedded costs, with provision for "stranded investment" costs. In addition, customers would be responsible for 3 percent or more in transmission losses, notwithstanding the fact that for a typical 100-mile, 345-kilovolt line, actual losses are 1 to 3 percent (Federal Energy Regulatory Commission 1989a, 60). The *nonfirm* transmission service rate would be capped at whichever was lower: the firm transmission service rate, or an allowance of 3 percent or more for transmission losses, plus one mill per kilowatt-hour, plus one-third of the savings in generating costs (after transmission losses) based on the difference between the marginal costs of the seller and buyer (Saacks 1992, Ex. JJS-2, 48–55).

Such transmission pricing would leave significant opportunities for the merged firm to exercise any increased market power created by the proposed merger. Under competition, both firm and nonfirm transmission rates could be substantially below the rates allowed under the proposed open access transmission tariff. Transmission rates under competition are determined by incremental, not embedded, costs. Where there is excess capacity, competitive transmission service rates may be no greater than actual transmission losses plus transactions costs. The incremental cost of a transmission transaction may even be negative: a transaction may cancel another in the opposite direction, reducing total losses.

Competition plays an important role in determining the prices of transmission service, including transmission service that is bundled with bulk power (see Chapter 3). Specifically, under its existing open access transmission tariff, if its merger with Gulf States Utilities were disapproved, Entergy's stand-alone nonfirm transmission service price would remain explicitly subject to negotiation and hence competition.

Aside from pricing, regulation of access to and expansion of transmission facilities under the proposed open access transmission conditions is unlikely to work well enough to prevent Entergy/Gulf States Utilities from exercising any increased market power. Regulation is imperfect because regulators have limited resources, including imperfect information about transmission capacities, costs, and effects on reliability. The merged firm could deny access to transmission or delay or deny requests for expansion of transmission facilities. Under the Energy Policy Act of 1992, the Federal Energy Regulatory Commission has issued a notice of proposed rulemaking to give potential users information concerning the availability of transmission. However, transmission-owning utilities have argued that the information provided would not be sufficient to allow third parties to evaluate whether there was sufficient transmission capacity to accommodate wheeling without threatening reliability (*Electric Utility Week,* June 28, 1993, 16–18). Furthermore, Entergy and Gulf States Utilities indicated in their merger application that the merged firm "shall be the sole judge of transmission service availability for a specified transaction" (Saacks 1992, Ex. JJS-2, 44).

Another problem with the provisions for expansion of transmission capacity is the requirement that wheeling service customers pay all incremental costs of capacity expansions. This ignores the benefits the host utility would enjoy from the expansion, such as increased reliability.

The terms of the open access transmission tariff have another problem that would limit the effectiveness of the tariff in constraining the exercise of market power. To be eligible for the open access transmission services, and thus to limit the exercise of market power, some transmission customers are required to pay an implicit price in addition to the wheeling charge. Utilities are required to offer comparable service on similar terms to the merged company and its affiliates, and customers may be required to forgo future purchases of power and energy at system average cost rates. This requirement may give the merged firm protection against competition in transmission.

RESTRICTIONS ON POSTMERGER PRICE INCREASES

In some cases, regulators have imposed restrictions on retail prices in approving mergers between electric utilities. Since merger applications are typically accompanied by claims that the proposed mergers will lead to substantial cost reductions, it has been common to impose ceilings on prices for a number of years in an effort to pass part of these savings through to customers. Of course, these restrictions on retail prices do not mitigate competitive problems relating to wholesale prices for power or transmission.

One specific issue that arises in mergers is the extent to which the merged company will be allowed to include the "merger premium"—that is, any excess of the purchase price over the regulatory book value of the acquired utility—in the rate base and recover it from future retail revenues. In light of the fact that electric utilities typically have unexercised market power in distribution, a policy that allowed acquiring firms to include the merger premium in the rate base could lead to monopoly pricing.

In the case of Kansas Power and Light's acquisition in 1992 of Kansas Gas and Electric, the Kansas Corporation Commission imposed a three-year rate freeze and some consumer rebates. The state commission allowed Kansas Power and Light to recover part of the merger premium by keeping 100 percent of initial "merger-related" savings, and to recover the remaining part of the merger premium by keeping 50 percent of additional merger-related savings. Of course, such a regulatory policy raises the issue of how savings that are "merger-related" are identified. In practice, the manner used by the Kansas commission to define merger-related savings (Robinson 1992) appears to include operating savings that would have been realized absent the merger.

CONCLUSION

When the federal antitrust agencies find that a merger is likely to have anticompetitive effects, it may be possible to defend the transaction based on evidence of substantial merger-specific cost savings or evidence that one of the merging firms would otherwise fail. In most cases, however, the alternatives are to sign a consent agreement containing provisions that remove the antitrust agencies' competitive concerns, to litigate, or to abandon the transaction. Consent agreements in horizontal merger cases normally rely on structural remedies to prevent a reduction in competition, generally through a divestiture of assets or licensing and supply agreements to facilitate new entry. The antitrust agencies generally avoid conditions that require continuing regulation of the merged firm.

By contrast, the Federal Energy Regulatory Commission's response to mergers that would reduce competition has generally been to approve the mergers based on transmission conditions that the Commission believes will prevent the exercise of increased market power in transmission and bulk power, as well as evidence of cost savings. The Commission's belief in the efficacy of transmission conditions is so great that it denied petitions to investigate competitive effects of the Entergy/Gulf States Utilities merger on the grounds that transmission conditions offered by the parties eliminated all competitive concerns.

The Commission's conclusion that regulation of transmission access and pricing can achieve the same results as competition is wrong. Utility mergers that reduce competition in transmission can be expected to lead to higher prices and to injure consumers, notwithstanding the Commission's efforts to prevent this result by regulation of the merged firm.

Our conclusion that transmission conditions are inadequate to compensate for a loss of competition between utilities does not imply that regulation of transmission access is inappropriate in other situations. Increased regulation of transmission access combined with more efficient pricing of transmission service can make markets for generation and supply of bulk power more competitive. This is one of the principles underlying the Energy Policy Act of 1992, which grants the Commission new powers to order wheeling service. Indeed, regulation of transmission access and pricing are likely to be sufficient to justify market-based pricing for bulk power in some situations.

NOTES

1. In the proposed merger between Southern California Edison and San Diego Gas and Electric, the Department of Justice was prepared to accept a reduction in competition in transmission combined with transmission conditions. "According to an informed source," however, "the Antitrust Division staff had wanted to oppose the merger in its entirety, but Assistant Attorney General James F. Rill disagreed, saying the transaction should be approved, but with conditions." *FTC:Watch*, Dec. 3, 1990, 6.

2. The Commission also justified its decision not to investigate the competitive effects of the merger on the grounds that no intervenor had demonstrated that *present* competition between the two systems is more than *de minimis*. This is not an appropriate standard. Intervenors demonstrated that Entergy's and Gulf States Utility's transmission systems offer alternative contract routes for bulk power between generators and customers. Even if Entergy and Gulf States Utilities did not both actually sell significant amounts of the same transmission service, an antitrust evaluation should consider whether the availability of a second, independent route constrains the pricing of the first.

9

Efficiency Analysis of Electric Utility Mergers

By Robert D. Stoner

The expanding role of competition in the electric power industry has increased not only the likelihood that mergers will have anticompetitive effects but also opportunities for mergers to have procompetitive effects. A merger may lead to efficiencies that enable a utility to reduce retail rates or to compete better in bulk power or transmission markets.

Efficiency effects should be weighed against anticompetitive effects by regulators deciding whether to approve a merger. Careful analysis is needed to assess the efficiency claims of merging utilities. Forecasts of cost savings are uncertain, and not every electric utility merger is likely to create efficiencies. For example, there is evidence that expansion of utilities beyond a certain size results in loss of managerial control and other diseconomies that are likely to outweigh any savings in cost. Moreover, many efficiencies may be achievable without merger.

This chapter defines efficiencies and discusses the proper methodology for measuring them. Characteristics of merging utilities that may produce significant efficiencies are identified. In particular, the relationship between utility size and costs is explored. The chapter also analyzes specific efficiency claims that have been made by merging utilities. The categories of efficiencies include fuel cost savings, deferred capacity savings, and nonfuel operation and maintenance cost savings. Examples are provided from the most important electric utility mergers of recent years. Areas of controversy in measuring efficiencies are highlighted.

PRINCIPLES OF EFFICIENCY ANALYSIS

For over a decade analysis of the efficiency claims made by merging parties has been a part of any merger investigation by the federal antitrust agencies. The approach described in the 1982 *Merger Guidelines* (U.S. Department of Justice 1982) was a significant change from antitrust enforcement policy of earlier decades, when there was a strong presumption to reject all efficiency claims (Stockum 1993). The 1968 *Merger Guidelines* stated that "[u]nless there are exceptional circumstances, the [Justice] Department will not accept as a justification for an acquisition normally subject to challenge under its horizontal merger standards the claim that the merger will produce economies" (U.S. Department of Justice 1968). Today, however, the antitrust authorities recognize that mergers may increase social welfare through the creation of efficiencies.

The 1992 *Horizontal Merger Guidelines* (U.S. Department of Justice and Federal Trade Commission 1992) explain the principles of efficiency analysis in a merger context. The first principle is that one should measure the extent to which a merger facilitates a more efficient use of resources, which results either when a given level of output is provided with fewer resources or when increased output is provided with the same or fewer resources. Mergers may promote costs savings in several ways, for example, by achieving economies of scale and reducing selling and other overhead expenses. Mergers, however, may also impose costs if it is more difficult to manage the resulting organization because of its large size or improper structure, or if problems cause maintenance and operating costs to increase. In addition, a proper accounting of cost savings must consider the difficulties that often occur in integrating the parties into one company.

A second principle of efficiency analysis is that the claimed savings should be merger-specific. This means that the savings could not be achieved by the parties acting independently or by contractual arrangements short of merger. Merging parties must explain why any savings produced by a merger could not be achieved by the firms standing alone. In many cases, stand-alone firms enter into contractual relationships that achieve many of the same savings that are available through merger.

However, economists have identified conditions under which contracting between independent firms may be difficult, and thus a merger may be necessary for cost savings. One such condition occurs when the parties have different information about the transaction, and the cost to a party of obtaining additional information is high. Contracts specify terms and obligations that will apply in the future. Differing views about the future may be another source of difficulty in contracting. Because of contracting difficulties some types of transactions are most efficiently handled within a single firm. It is easy, however, to overstate the difficulties of writing a contract. Independent businesses regularly enter into both long-term and short-term contracts. The reason is that the alternative to a contract—common ownership—presents difficulties of its own.

The principles that the federal antitrust authorities use in analyzing efficiencies have not been adopted fully by the Federal Energy Regulatory Commission. In particular, there has been a great deal of controversy about whether efficiencies that could be achieved without the merger should still be "counted" as merger efficiencies (Federal Energy Regulatory Commission 1991a, 61,994–95). Several recent decisions by the Commission have explicitly departed from antitrust agency standards and declared that even cost reductions that are not merger-specific should still be viewed as efficiencies attributable to the merger (Federal Energy Regulatory Commission 1991a, 61,994–95). Nonetheless, Federal Energy Regulatory Commission proceedings and decisions have paid enough lip service to the antitrust standards for measuring efficiencies that the presentations of merging and intervening parties invariably attempt to separate merger-specific cost savings from savings available on a "stand-alone" basis.

Calculation of Merger-Specific Cost Savings

To calculate merger efficiencies, one must follow four steps. First, estimate the costs for each company assuming that it remains independent. This is generally called the "stand-alone" case. Stand-alone costs provide the baseline from which to estimate merger efficiencies. Stand-alone costs should be estimated assuming that both firms take all prudent and reasonable steps to operate efficiently, including entering into cost-reducing contracts and trades. If stand-alone costs do not include all such prudent and reasonable actions, the stand-alone costs will be too high and projected efficiencies will be overstated. Second, estimate the costs for the proposed combined company. As with stand-alone costs, the combined costs are estimated assuming the merged firm takes prudent and reasonable steps to operate efficiently. Third, calculate the difference between stand-alone costs and combined costs. If stand-alone costs are greater than combined costs, cost savings result from the merger. If stand-alone costs are less than combined costs, the merger raises costs. Finally, calculate the present worth of efficiencies by taking the present discounted value of any future cost savings.

Estimated merger savings that would occur over a number of years should be expressed in terms of their *present discounted value.* Quite apart from inflation, future cost savings are not worth as much as the same cost savings today. A dollar saved today could be invested and earn interest. Hence it is worth more than a dollar saved in a future year. Cost savings in each future year must be discounted (effectively, multiplied by a number smaller than one) to reflect the relative values of a dollar of savings today and in future years. The farther in the future that anticipated savings would occur and the higher the rate used to discount future savings, the smaller will be the present discounted value of a given level of future savings, and hence the less will be the weight that should be given to that savings.

In addition, discounting can be used to account for the fact that merger savings are subject to risk and may not be achieved (Baumol 1977, 619–25). Just as financial assets with high risk (for example, junk bonds) carry high interest rates in the market, a relatively high discount rate can be used to adjust for the risk that cost savings will not materialize. Discount rates somewhat higher than a utility's cost of capital are not unreasonable, since there is greater risk inherent in estimates of cost savings than in a utility's regulated profit stream (Stoner 1993, 21).

Reasons for Miscalculation of Cost Savings

There has not been a study of the *ex post* success or failure of electric utility mergers relative to *ex ante* predictions. However, there is a large literature on the success or failure of mergers generally in a wide range of industries. Before a merger is consummated the merging parties anticipate cost savings and efficiencies (Jensen and Ruback 1983; Bradley et al. 1988; Dewing 1921). Mergers, however, rarely lead to the lower costs, increases in market shares, or increased profits that the merging parties expect (Ravenscraft and Scherer 1989; Mueller 1985; Louis 1982; Dewing 1921).

Merging firms often do not achieve the anticipated cost savings for a number of reasons. One reason is loss of management control. A substantial literature describes the limits of managerial control as firm size increases. Williamson (1975, chaps. 2 and 7) stresses that expansion of the enterprise usually requires the addition of hierarchical layers. These additional management layers often result in greater costs and less efficient production. There is also evidence that managers of acquired firms are replaced more rapidly than managers in general (Hartman 1990a, 22). This management turnover leads to a loss of management control, which leads in turn to inefficiencies and lower profits. Ravenscraft and Scherer (1989, 115) state that the reason "for acquired units' sharp profit decline must be control loss owing to more complex organizational structures and lessened managerial competence and/or motivation."

Managerial empire building also can frustrate achievement of anticipated cost savings. Managers may have incentives to expand firm size beyond the efficient production levels (Baumol 1977, chap. 15). This motivates managers to try to justify mergers with overly optimistic *ex ante* savings projections. Empire building has been suggested as a motive in an electric utility merger (Owen 1989).

Unanticipated operating and integration problems is a third reason why cost savings may not materialize. Problems that delay cost savings for several years can significantly reduce the present discounted value of merger benefits, since more distant future benefits are more heavily discounted.

SOURCES OF EFFICIENCIES FOR ELECTRIC UTILITIES

Coordination of utility system operations, either within a firm or between firms through contract or pooling, may permit several types of economies to be achieved. These economies are largely "vertical" in the sense that they occur through the coordination of generation and load. One of the most difficult aspects of the efficiency analysis of mergers is to differentiate those integration economies that can best be achieved within one firm and those that are available through alternative means short of merger.

1. *Systemwide Scale Economies.* A larger coordinated system allows the consolidation of spatially dispersed loads so that they can be served by a relatively small number of large generating plants rather than by a large number of very small plants.

2. *Dispatch Economies.* Economic dispatch permits the least-cost choice of generating units to meet particular load characteristics. Such economies can be exploited only if there is a broad enough mix of plants featuring baseload, intermediate load, and peaking capabilities.

3. *Diversity Economies.* To the extent that load patterns differ across areas or customer types, there can be economies in both the construction and operation of generating capacity if these areas or customer types are combined. The most obvious example is where areas peak in different seasons of the year or at different times of the day. Aggregation of loads can reduce the joint system's need for capacity below the sum of the capacities needed by the two utilities taken alone, and aggregation

of loads may permit lower cost operation of a given set of generation units. Another aspect of diversity that may be exploited to lower costs is differences in the predominant type of customer between systems (for example, residential, commercial, industrial, and wholesale). Diversity economies may also exist if generation fuel types differ across utilities. In that case, combined operation may result in lower operating costs because of greater use of plants with low fuel costs.

4. *Density Economies*. Unit costs may decline as more customers are served by a given utility system, or as consumption per customer increases. Density economies do not occur when a utility simply expands its service territory with no change in either the number of customers within a given service area or the amount of electricity purchased per customer (Roberts 1986).

5. *Maintenance Economies*. Maintenance of generating units must be performed on a regular basis, and nuclear units must be shut down periodically for refueling. When such maintenance is performed, higher cost replacement power may be required. Coordination of maintenance scheduling across numerous interconnected plants serving a common load may reduce the need for costly replacement power.

6. *Reliability and Emergency Economies*. Utilities need reserve generating and transmission capacity because they serve uncertain loads and because generating units and transmission facilities are subject to failure. Coordination of operations can soften the impact of these uncertainties, necessitating lesser reserves for the same level of reliability. Coordinated operation of interconnected generating plants also improves the ability of a system to avoid load losses during emergencies such as forced outages.

Mergers between electric utilities are more likely to produce these types of vertical and other efficiencies under certain conditions. Physical interconnection between merging utilities is necessary to achieve many of these economies. Diversity economies depend on the extent of diversity between the merging utility systems' load patterns or fuel types. Mergers that simply expand service territories without increasing the intensity of use of the system will not necessarily produce economies. In addition, consolidation short of merger will allow many of these efficiencies to be realized.

Generating Unit, Plant, and Firm Size Economies

The focus of improved performance in the electric power industry has changed over time.[1] Initially, transmission technology was relatively undeveloped, service territories were limited in size, and the industry consisted of small isolated plants generating in localized areas. As the electric power industry grew, the major gains came from lower generating costs resulting from economies for larger generating units and plants. This process reached its zenith in the 1950s and 1960s. Optimal firm size, in turn, was determined by the efficient array of generating plants necessary to serve local baseload, intermediate load, and peaking requirements for power.

Since the late 1960s the predominance of scale economies in generation at the unit and plant level has subsided. There has been increasing evidence of diseconomies at the largest unit and plant sizes for both fossil and nuclear units. Several studies have found that economies of scale for baseload fossil fuel units are exhausted at about 400 to 500 megawatts because any further economies are offset by the need for higher reserve margins and the lack of unit flexibility (Loose and Flaim 1980; Schroeder et al. 1981; see also Joskow and Schmalensee 1983). Optimal size for baseload fossil fuel plants appears to be about 800 megawatts, or two optimal units. A similar limitation of unit and plant scale economies, at about 1,000 megawatts and 2,000 megawatts, respectively, has been found for nuclear units (Behrens 1985; see also Joskow and Schmalensee 1983). The limitation on unit and plant scale economies has been heightened by inflationary pressures, which have raised the cost of capital, and environmental regulations, which have increased the logistical difficulty of constructing large plants (Joskow 1974). Scale economies for intermediate and peak-load generating facilities are even more limited.

As the importance of *plant-level* scale economies in generation subsided, however, technological changes in the other vertical stages of utility operation increased *firm-level* economies. Improvements in transmission technology reduced line losses and permitted the connection of formerly isolated generating plants into larger operating systems. The advent of computerization in conjunction with transmission improvements increased the opportunities for coordination across the entire vertical span of utility operations. This allowed the optimization of generation to the particular load characteristics of a utility. Some of these economies were best exploited within a single utility. Some also could be achieved through power pools and other contractual relationships that did not necessitate increases in firm size.

There is an extensive literature on the relationship between utility scale and unit generating costs that attempts to measure the extent of one type of firm-level economies. This literature can shed light on which utility mergers are likely to lead to cost savings and which are not. Statistical studies over a twenty-five-year period have attempted to determine how utility production costs vary with firm size, holding constant other important determinants of cost such as fuel mix, capacity utilization, participation in power pools, and customer mix (Hartman 1990a; Atkinson and Halvorsen 1984; Mayo 1984; Christensen and Green 1976, 1978; Huettner and Landon 1978). We can summarize this literature by describing the ranges of firm size over which average production costs decline, remain constant, and increase. Size ranges are expressed here in gigawatt-hours of output, regardless of whether the study measured size in terms of capacity (megawatts or gigawatts) or output (gigawatt-hours). Conversion between capacity and output measures was accomplished assuming 8,760 hours per year and using the national average load factor of 55 percent.

The evidence indicates that production costs are likely to decline as the firm's production increases up to some point between 12,000 and 30,000 gigawatt-hours per year. For illustrative purposes, utilities the size of Wisconsin Public Service and Oklahoma Gas and Electric fall within this range. Beyond some point in this range, greater production tends not to produce additional unit cost savings. The range of relatively constant average unit costs continues up to approximately 54,000 to 67,000 gigawatt-hours per year. Again for illustrative purposes, utilities the size of Duke Power in North Carolina and Virginia Electric Power fall into this range. Beyond some point in this range, loss of managerial control and other size diseconomies outweigh any cost savings related to firm size, and unit production costs tend to rise. Utilities such as Commonwealth Edison in Illinois and American Electric Power are in this range. Summarizing this literature, Figure 9.1 shows the ranges within which average costs decline, remain constant, and increase.

These studies of the relationship between generating costs and utility size give an initial indication of which mergers are likely to reduce generating cost, which mergers are likely to yield little or no savings, and which mergers are likely to increase generating costs. In general, mergers among small utilities (for example, where the merged utility will generate significantly less than 30,000 gigawatt-hours per year) are likely to produce cost savings. It would make sense to presume that these mergers will produce cost savings in generation. Claims of generating cost savings relating to firm size for mergers creating utilities producing more than 30,000 gigawatt-hours per year should be viewed more skeptically. Merged entities in this range may or may not produce savings, depending on the nature of the merger. For example, a merger of three firms, each producing 12,000 gigawatt-hours per year, would be likely to produce savings even though the total postmerger production (36,000 gigawatt-hours) is greater than 30,000 gigawatt-hours. By contrast, a merger of two firms each producing 25,000 gigawatt-hours per year would be less likely to produce savings. Finally, mergers creating firms with production beyond some point in the 54,000 to 67,000 gigawatt-hours per year range are unlikely to produce generating cost savings relating to firm size. Such firms have higher costs than smaller firms, other things equal. Firms proposing mergers that would result in utilities producing more than 30,000 gigawatt-hours per year should bear the burden of providing persuasive evidence for their *ex ante* estimates of generating cost savings.

Figure 9.1: Average Cost of Power Generation

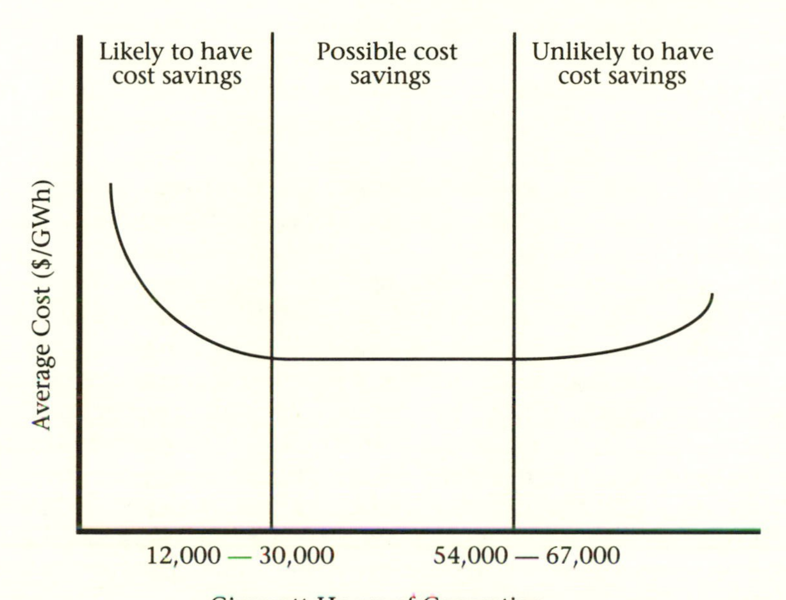

Use of the combined output of the merged firm as an indicator of the extent of likely merger savings relating to generating costs is subject to three important caveats. First, the studies underlying Figure 9.1 relate to the costs of generation, not to all categories of utility costs. The effect of the merger on other categories of costs would need to be evaluated as well. Below we review studies that deal with the relationship between overhead costs and utility size.

Second, the studies underlying Figure 9.1 examine the effect of utility size on unit generating costs *holding a number of variables constant.* However, a utility merger may change not only the scale of the utility but also some of those other variables. For example, load diversity, fuel mix, and capacity utilization may be affected by the merger. In that case, the effect of a merger on generating costs may be more favorable than Figure 9.1 would suggest.

Third, Figure 9.1 may overstate the advantages of large firm size to the extent that some of the generating costs savings that large firms in the study samples achieved would be available through contracts or power pools. Modern electric utility systems are extensively interconnected, engage in pooling of various degrees of tightness, and buy and sell power and transmission service through contracts. Some of the benefits of large firm size are likely to be increasingly available through pooling and contracting. Statistical cost studies of firm size economies have made only limited attempts to control for this type of effect, for example, by including "the extent of pooling" as an additional variable explaining costs. Thus it is unlikely that the studies fully control for the ability of utilities to achieve "scale" benefits by means other than increasing firm size.

Figure 9.1 relates to generating costs. Several studies relate the size of electric utilities to overhead costs. These studies similarly indicate that per unit overhead costs do not decline and may actually increase as firms increase above a certain size. Based on a sample of seventy-four electric utilities, Huettner and Landon (1978) found that administrative and general expenses per gigawatt-hour increased with firm size beyond 12,000 gigawatt-hours per year and that customer account expenses per gigawatt-hour increased with firm size beyond 8,190 gigawatt-hours per year. Hartman (1990b, II-21–22) assessed the relationship between labor overhead costs and scale using a sample of 181 utilities. He found that optimal firm size was in the neighborhood of 46,000 gigawatt-hours per year; above that size unit costs rose. Finally, Harris performed a more limited study based on a sample of 25 electric utilities serving 750,000 or more customers. He found that administrative and general expenses per customer did not decline with greater firm size (Harris 1989, exhibits BCH-20, 21, and 26). These empirical results are consistent with the work of Williamson and others, contending that large firms' loss of management control can offset theoretical gains in production efficiency.

One possible methodology, based on the virtually unanimous findings of the above studies, is to develop "presumptions" that can aid regulators in determining when efficiencies relating to firm size are likely to materialize. The establishment of presumptions about whether an electric utility merger is likely to yield savings relating to firm size would be similar to the rebuttable presumptions regarding anticompetitive effects based on concentration indices articulated by the 1992 *Horizontal Merger Guidelines*. Of course, any presumptions relating to efficiencies and firm size should be rebuttable based on persuasive merger-specific evidence.

EFFICIENCY CLAIMS IN SPECIFIC UTILITY MERGERS

Table 9.1 shows the savings claims made by the parties in connection with a number of electric utility mergers since 1967. The claimed savings from each merger are expressed as a percent of the annual operating revenues at the time of the merger. These ratios vary from less than 1 percent to 5.4 percent.

An important reason for the differences in the magnitude of savings is that some mergers have little opportunity for substantial vertical economies of the types discussed above. For these mergers, the largest source of efficiencies is likely to be in overhead consolidation or other manpower savings—often the most speculative and unverifiable type of merger efficiency claim.

More detailed information on efficiencies is available concerning five large electric utility mergers or proposals since 1988. The specific efficiency arguments in each of these cases and the manner in which efficiencies were treated by regulators are described in this section.

Utah Power and Light/Pacific Power and Light

The parties in the merger of Utah Power and Light and Pacific Power and Light predicted savings of 5.4 percent of annual operating revenue, the largest savings relative to size of operations of the mergers in Table 9.1. The large savings were mainly the result of differences between the two systems. Pacific Power and Light has a winter peak, and Utah Power and Light has a summer peak. As a result, the parties claimed that several types of diversity-related economies were available to the merged firm. First, lower cost generating plants could be run more often because the noncoincident seasonal peaks of the combined utility were "smoother" than the seasonal peaks of each of the companies standing alone. Second, capacity additions could be deferred because of the lower combined peak and resulting lower reserve requirements. In addition to these potential savings, the merger applicants claimed that Pacific Power and Light had surplus power that would enable Utah Power and Light to defer capacity additions on its system. Finally, the parties claimed that economies of scale would allow operating savings by combining administrative functions, consolidating inventories, and sharing services between the operating divisions. The applicants

Table 9.1

Utility Merger Savings Claimed by Applicants

Utilities	Approved	Annual Merger Savings	
		$ Mil.	% of Electric Oper.Rev.
Eastern Gas & Fuel, Boston Gas, Brockton Gas	1967	$.29– $.36	<1.0%
Hawaiian Elec., Hilo Elec.	1970	$.21	<1.0%
American Elec. Power, Columbus and Ohio Elec.	1978	$49.5	1.5%– 2.6%
Centerior Energy, Cleveland Elec. Illum., Toledo Edison	1986	$18.9	1.0%
Southern Company, Savannah Elec. Power	1988	$50	<1.0%
Utah Power and Light, Pacific Power and Light	1988	$113	5.4%
Southern California Edison, San Diego Gas and Elec.	rejected 1991	$142	2.0%
Eastern Utility Associates, Unitil, Fitchburg*	1990	$8.5– $13.2	1.8%– 2.8%
Kansas P & L, Kansas G &E*	1991	$25.8	2.5%
Northeast Utilities, Public Service of New Hampshire	1992	$120	3.7%
Entergy, Gulf States Utilities	1993	$170	2.8%

Source: Adapted, with additional entries, from Hartman 1990a and Arthur D. Little 1990.

*Takeover

projected merger-related savings of approximately $505 million over the five-year period following the merger.

The Federal Energy Regulatory Commission administrative law judge rejected most of the diversity-related benefits the applicants had claimed because the benefits were deemed attainable absent the merger. The Commission reversed the judge's decision, stating that "the possibility of achieving a particular benefit through a contractual arrangement does not diminish the cost savings associated with that benefit" (Federal Energy Regulatory Commission 1988, 61,299). The Commission, however, rejected for lack of substantiation a major portion of the claimed savings from combining administrative functions. Eventually, the merger was approved with conditions to ameliorate potential anticompetitive effects. A major reason that the merger was approved was the recognition that significant efficiencies were attainable.

Southern California Edison/San Diego Gas and Electric

The annual savings projected by the merging parties in the California utility merger were higher than in the Utah Power and Light case ($142 million per year compared with $113 million per year). The claimed savings as a percentage of annual operating revenues, however, were smaller (2 percent compared with 5.4 percent). The parties in the California merger claimed three principal kinds of savings: resource deferral savings of $37 million annually stemming from the use of Edison's existing generation and transmission capacity to meet San Diego Gas and Electric's near-term capacity requirements; transmission service savings of $27 million annually from a claimed increased ability of the combined system to use existing transmission systems to the Pacific Northwest and Southwest; and labor savings related to corporate staff reductions of $105 million annually. There are also some expected merger costs. Total projected net savings over the 1990–2000 period were approximately $1.7 billion.

The Federal Energy Regulatory Commission administrative law judge and the California Public Utilities Commission rejected almost all of the merger benefits claimed by the parties (Federal Energy Regulatory Commission 1990e, 65,125; California Public Utilities Commission 1991b, 10–15). Labor cost savings, viewed as subjective and unverifiable, were considered to be partly available absent the merger and likely to be offset to some extent by diseconomies of large size. Capacity deferral benefits were found to be exaggerated because the period of analysis was cut off before significant costs were experienced. The Commission believed the capacity deferral benefits would be available absent the merger and, in any case, would not be long-term because of the similarity of the peak demands of the two utilities. Because both California utilities have summer peaks, the total amount of capacity needed to serve the peak demand of the merged company was found to be identical to that required for the stand-alone systems (Federal Energy Regulatory Commission 1990e, 65,134). The transmission service savings were also found to be available absent the merger.

The federal administrative law judge stated that the Commission's finding in *Utah Power and Light* that merger benefits should be counted even if they could be attained absent the merger was "not applicable precedent" (Federal Energy Regulatory Commission 1990e, 65,119). Both the federal administrative law judge and the California commission found anticompetitive effects that they thought would be significant and likely. These anticompetitive effects and the efficiency findings caused them to reject the merger.

Northeast Utilities/Public Service of New Hampshire

Northeast Utilities in Connecticut and Massachusetts claimed that the most important merger benefit was the resolution of Public Service of New Hampshire's bankruptcy. No specific monetary value was claimed for this benefit. Northeast Utilities also claimed benefits of $364 million over ten years from combining the two utilities into "a single participant" in NEPOOL, the New England power pool. Other savings claimed by the merging parties over a ten-year period included $527 million in operating and maintenance costs from Northeast Utilities' operation of Public Service of New Hampshire's Seabrook nuclear facility; $100 million from improved availability of Public Service of New Hampshire's fossil steam generating units; $124 million from reduced administrative and general expenses; and $39 million from reduced coal purchasing costs for Public Service of New Hampshire (Federal Energy Regulatory Commission 1991a, 61,994). Claimed benefits totaled approximately $1.2 million over a ten-year period. An important reason for lower fuel costs under the merger is that demand on the Northeast Utilities system peaks in summer and demand on the Public Service of New Hampshire system peaks in the winter. This difference was expected to allow the combined system to operate more efficiently than could the stand-alone companies (Federal Energy Regulatory Commission 1991a, 61,988).

While the Federal Energy Regulatory Commission disputed the NEPOOL savings, it accepted, with little comment, most of the other merger benefits that were claimed. The Commission allowed the merger subject to certain transmission conditions. Without conditions, "the merger would produce substantial benefits but . . . these benefits would be outweighed by the merger's likely anticompetitive effects" (Federal Energy Regulatory Commission 1992a, 61,195).

Kansas Power and Light/Kansas Gas and Electric

Efficiencies claimed by Kansas Power and Light and Kansas Gas and Electric were considerably smaller than in the other mergers considered here, totaling approximately $140 million over five years. Claimed savings over this period came from three main areas: labor savings from consolidation of field office and head office functions, $101.3 million; production savings from centralized dispatch, planned maintenance, and more efficient fuel use, $24.5 million; and avoided expenditures on computer systems, $17.5 million. Costs to attain these savings were put at $11 million (Hartman et al. 1991).

After conducting an in-depth study of the claimed efficiencies, the Missouri Public Service Commission apparently found that the efficiency claims were exaggerated and the costs of the acquisition were underestimated (Hartman et al. 1991, 58–59). The Federal Energy Regulatory Commission allowed the merger to proceed by accepting an uncontested settlement without scrutinizing the claimed efficiencies (Federal Energy Regulatory Commission 1991b).

Entergy/Gulf States Utilities

Annual merger savings projected by Entergy and Gulf States Utilities were of similar magnitude to those in the California utility merger, with total projected savings over the period 1994 to 2003 of $1.7 billion. The main categories of projected savings were fuel and purchase power savings, $84.9 million annually; deferred capacity savings, $18.4 million annually; and operating and maintenance cost savings, $67.3 million annually (Stoner 1993). The relatively large magnitude of projected fuel savings is attributable to diversity between the two systems in *fuel type*. Gulf States Utilities is more dependent than Entergy on relatively expensive natural gas generation. It was projected that the merger would permit replacement of natural gas generation by Gulf States Utilities with coal generation by Entergy and with off-system power purchases by the combined system. The relatively small level of deferred capacity savings is attributable to the similar load characteristics of the Entergy and Gulf States Utilities systems. Both are summer-peaking utilities, and both peak at the same time of day in the summer (late afternoon) and the winter (early morning). The federal administrative law judge accepted only the likelihood of significant fuel and purchase power savings, stating that deferred capacity savings were unlikely and operating and maintenance cost savings had not been convincingly demonstrated (Federal Energy Regulatory Commission 1993d, 62, 301). The administrative law judge's decision is currently being reviewed by the Commission.

Common Flaws in Efficiency Claims

The efficiency claims made by the merging parties in the above proceedings have some common flaws. First, the merging parties in several of the proceedings have stated that there is no need to discount the future stream of savings because, in effect, "when savings accrue to ratepayers they will not be in discounted dollars." This is clearly a fallacious argument. Without discounting there is no way to put savings that are slated to occur in different periods on the same "footing." Cost savings tomorrow are not worth as much as cost savings today. In addition, discounting can be used as a means to take account of the increased riskiness of distant projections compared to near-term estimates.

Second, as a rule the stand-alone case has been incorrectly specified. Stand-alone costs should be estimated assuming that both merging firms take all prudent and reasonable steps to operate efficiently, including entering into cost-reducing contractual relationships with each other or third parties. If stand-alone costs are not calculated taking into account such prudent and reasonable actions, then stand-alone costs will be too high and projected efficiencies will be overstated. In general, merger applicants have made little effort to specify the stand-alone case correctly.

For example, in the Entergy/Gulf States Utilities merger, very large fuel cost savings to Gulf States Utilities were forecast based on the assumption that, without the merger, rising natural gas prices (relative to coal prices) would hurt gas-intensive Gulf States Utilities. The merger was postulated to solve this problem because Entergy has excess coal-based energy. Yet there was no attempt to determine what Gulf States Utilities' optimal stand-alone reaction to rising gas prices would have been. Presumably, Gulf States Utilities would have been spurred to seek cheaper coal energy in the stand-alone case, either from Entergy or others. By contrast, in the merging parties' analysis, Gulf States Utilities was simply assumed to react passively and continue past practices.

Third, claimed benefits deriving from capacity deferral have been flawed in a number of the proceedings. Capacity deferrals do not refer to permanently lower capacity requirements but simply to delays in adding to capacity or engaging in demand-reducing programs. Merging parties typically estimate merger benefits and costs over a limited number of years. If the merging parties are able to claim that a merger would permit a delay in certain capital costs beyond this measurement period, they sometimes claim the entire amount of the delayed capital cost as a merger benefit—ignoring the fact that the addition to capacity would still have to be made after the end of the measurement period. The result is to exaggerate merger benefits. This effect is accentuated because capacity deferral often results in increases in fuel costs as more efficient units are postponed. This can have fuel cost implications that extend beyond the measured period.

Fourth, the claims of savings in customer service and overhead costs have tended to be asserted rather than demonstrated. The methodology has generally been to identify a "cost gap" between the acquiring and acquired firm and then to assume that the merger will allow the combined firm to achieve the costs of the lowest-cost party. Alternatively, a cost benchmark, based on the lowest cost utility performers across the country, is adopted, and then simply applied to the merged firm. These methodologies are unreliable. They do not identify the reasons for the "cost gap" or inferior performance of the premerger entity, and they do not specify how the merger will bridge the cost gap. For example, if the reason for the premerger cost gap between the parties is that there are differences in the composition of customers or in labor costs, the merger will not necessarily result in lower overall costs. Furthermore, if there are diseconomies of large size, as the statistical cost literature projects, expected overhead and administrative cost savings may be offset by higher bureaucratic or other costs.

CONCLUSION

Under certain circumstances, utility mergers can produce substantial cost savings. Mergers between relatively small utilities, or between utilities with different load patterns or fuel types, may produce significant savings. If these savings are not available without the merger, they are important counterweights to anticompetitive effects in an antitrust review.

Under other conditions, however, cost savings are dubious and should be viewed skeptically by the regulatory authorities. Scale efficiencies are unlikely in the case of mergers among very large utilities. Furthermore, in a broad cross-section of industries, before-the-fact estimates of merger economies have been greatly exaggerated.

Efficiency claims in a number of recent electric utility mergers have been very large. Except in a few cases, many of the claimed savings have been rejected by the regulatory authorities. Claims of deferred capacity savings and administrative overhead savings have been especially problematic.

NOTE

1. This section borrows significantly from Hartman 1990a and Joskow and Schmalensee 1983.

Bibliography

Acton, Jan Paul and Stanley M. Besen. 1985. *The economics of bulk power exchanges*. Santa Monica, Calif.: Rand Corp. N-2277-DOE.

Ahern, William R. 1985. *Position of the California Public Utilities Commission's Public Staff Division on the regulation of utility diversification in California*. California Public Utilities Commission. San Francisco. Available as Federal Energy Regulatory Commission, Docket No. EC89-5-000, Southern California Edison and San Diego Gas and Electric, Exhibit 734. Washington, D.C.

Alger, Dan and Susan Braman. 1993. Competitive joint ventures and reducing the regulation of natural monopolies. Economists Incorporated and Federal Trade Commission. Washington, D.C. Mimeo.

Aronson, Clifford H. and James A. Keyte. 1992. Cutting the suit to fit the cloth: Innovative solutions to merger challenges by the DOJ and FTC. *Antitrust* (Summer): 26–30.

Arthur D. Little. 1990. Evaluation of EUA's proposed acquisition of UNITIL and Fitchburg: Report to Gaston and Snow.

Atkinson, Scott E. and Robert Halvorsen. 1984. Parametric efficiency tests, economies of scale, and input demand in U.S. electric power generation. *International Economic Review* 25 (October): 647–62.

Averch, Harvey and Leland L. Johnson. 1962. Behavior of the firm under regulatory constraint. *American Economic Review* 52 (December): 1052–69.

Baker, Jonathan B. 1987. Why price correlations do not define antitrust markets: On econometric algorithms for market definition. Federal Trade Commission, Bureau of Economics, Working Paper 149. Washington, D.C.

Baughcum, M. Alan. 1989. *Prepared direct testimony*. Federal Energy Regulatory Commission, Docket EC89-5-000, Southern California Edison and San Diego Gas and Electric, Exhibit 809. Washington, D.C.

Baumol, W. 1977. *Economic theory and operations analysis*. London: Prentice-Hall.

Baxter, William F. 1983. Conditions creating antitrust concern with vertical integration by regulated industries—"For whom the Bell doctrine tolls." *Antitrust Law Journal* 52: 243–47.

Bayless, Charles E. 1992. Transmission pricing: Striking a balance. *Public Utilities Fortnightly* 130 (October 15): 13–17.

Behrens, Carl. 1985. Small nuclear power plants: Financing ease may balance scaling factor. *Energy Policy* (UK) 13 (August): 360–70.

Beierlein, James G., James W. Dunn, and James C. McConnon, Jr. 1981. The demand for electricity and natural gas in the northeastern United States. *Review of Economics and Statistics* 63 (August): 403–8.

Blattenberger, Gail R., Lester D. Taylor, and Robert K. Rennhack. 1983. Natural gas availability and the residential demand for energy. *Energy Journal* 4 (January): 23–45.

Boettcher, D. 1990. *Deposition.* California Public Utilities Commission, App. 88-12-035, Southern California Edison and San Diego Gas and Electric. San Francisco.

Bradley, Michael, Anand Desai and E. Hahn Kim. 1988. Synergistic gains from corporate acquisition and their division between stockholders of target and acquiring firms. *Journal of Financial Economics* 21: 3–40.

Braeutigam, R. R. and J.C. Panzar. 1989. Diversification incentives under "price-based" and "cost-based" regulation. *Rand Journal of Economics* 20 (Autumn): 373–91.

Braman, Susan. 1992. *Theory and applications of competitive joint ventures*. Ph.D. diss., Georgetown University, chap. 6. Washington, D.C.

———. 1993. Competitive joint ventures and electricity transmission. Georgetown University. Washington, D.C. Mimeo.

Brandon, Paul S. 1971. The electric side of combination gas-electric utilities. *Bell Journal of Economics and Management Science* 2 (Autumn): 688–703.

Brennan, Timothy J. 1987. Why regulated firms should be kept out of unregulated markets: Understanding the divestiture in *United States v. AT&T*. *Antitrust Bulletin* 32 (Fall): 741–93.

———. 1990. Cross-subsidization and cost misallocation by regulated monopolists. *Journal of Regulatory Economics* 2 (March): 37–51.

California Public Utilities Commission. 1986a. *Audit report on Pacific Telesis, Inc.* Washington, D.C.: National Association of Regulatory Utility Commissioners.

———. 1986b. *Decision 86-07-004*. San Francisco.

———. 1987. *Decision 87-05-060*. San Francisco.

———. 1990. *Decision 90-09-088*. San Francisco.

———. 1991a. *Proposed decision of administrative law judges Carew and Cragg. App*. 88-12-035. Southern California Edison and San Diego Gas and Electric. San Francisco.

———. 1991b. *Decision 91-05-028*. App. 88-12-035. Southern California Edison and San Diego Gas and Electric. San Francisco.

———. 1991c. *Decision 91-12-076*. App. 90-12-018. Southern California Edison. San Francisco.

California Public Utilities Commission, Division of Ratepayer Advocates. 1987. *Second petition of the Division of Ratepayer Advocates to dismiss Edison's application . . . ,* California Public Utilities Commission, App. 85-12-012. San Francisco. Available as Federal Energy Regulatory Commission, Docket No. EC89-5-000, Southern California Edison and San Diego Gas and Electric, Exhibit 737. Washington, D.C.

———. 1988. *Report on the reasonableness of Southern California Edison non-standard power purchase contracts with qualifying facilities.* California Public Utility Commission, App. 88-02-016, Southern California Edison reasonableness review. San Francisco.

———. 1989. *Concurrent opening brief.* California Public Utilities Commission. App. 88-02-016, Southern California Edison reasonableness review. San Francisco.

Calvani, Terry. 1987. Statement in *The Federal Trade Commission's agenda in the energy industry*. American Bar Association, Section on Natural Resource Law, June 2. Washington, D.C.

Carr, Ronald G. 1982. The antitrust division perspective: Mergers and acquisitions in the natural gas industry. Paper presented to American Gas Association, March 31.

Christensen, Laurits R. and William H. Green. 1976. Economies of scale in U.S. electric power generation. *Journal of Political Economy* 84: 656–76.

———. 1978. An econometric assessment of cost savings from coordination in U.S. electric power generation. *Land Economics* 54 (May): 139–55.

Cicchetti, C. and W. Hogan. 1988. Including unbundled demand side options in electric utility bidding programs. Harvard University, Energy and Environmental Policy Center. Discussion Paper. Cambridge, Mass.

Coate, Malcolm B. 1993. Merger analysis in the courts. Federal Trade Commission, Bureau of Economics, Working Paper. Washington, D.C.

Coate, Malcolm B., Andrew N. Kleit, and Rene Bustamante. 1993. Fight, fold or settle?: Modeling the reaction to FTC merger challenges. Federal Trade Commission, Bureau of Economics, Working Paper No. 200. Washington, D.C.

Coate, Malcolm B. and Fred S. McChesney. 1992. Empirical evidence on FTC enforcement of the *Merger Guidelines*. *Economic Inquiry* 30 (April): 277–93.

Cohen, Barry N., John A. Holte, and Paul J. Werbos. 1985. *Demand analysis system elasticities*. Washington, D.C.: Energy Information Administration. DOE/EIA-0475.

Collins, Wayne D. and James R. Loftis III. 1988. *Non-horizontal mergers: Law and policy.* Washington, D.C.: American Bar Association, Section of Antitrust Law. Monograph 14.

Collins, William H. 1973. Comparative performance of combinations and separately managed electric utilities. *Southern Economic Journal* 40 (July): 80–89.

Congressional Research Service. 1991. *Electricity: A new regulatory order?* U.S. House of Representatives, Comm. on Energy and Commerce. Washington, D.C.

Cross, Phillip S. 1992. Cogeneration: Growing risk in a complex market. *Public Utilities Fortnightly* 130 (Dec. 1): 39–42.

Demsetz, Harold. 1968. Why regulate utilities? *Journal of Law and Economics* 11 (April): 55–65.

Dewing, A. S. 1921. A statistical test of the success of consolidations. *Quarterly Journal of Economics* 36: 84–101.

Earley, Wilbur C. 1984. FERC regulation of bulk power coordination transactions. Federal Energy Regulatory Commission, Office of Regulatory Analysis, Staff Working Paper. Washington, D.C.

Edison Electric Institute. Annual. *Capacity and generation of non-utility sources of energy*. Washington, D.C.

———. Annual. *Statistical yearbook of the electric utility industry.* Washington, D.C.

Elston, Paul J. 1991. *Testimony on behalf of National Independent Energy Producers.* U.S. Senate, Comm. on Banking, Housing, and Urban Affairs, Subcomm. on Securities, Hearings on Title XV of S. 1220, September 17. Washington, D.C.

Energy Information Administration. 1983. *Interutility bulk power transactions.* Washington, D.C. DOE/EIA-0418.

———. 1985. *Financial statistics of selected electric utilities 1983.* Washington, D.C. DOE/EIA-0437(83).

———. 1991. *Inventory of power plants in the United States 1990.* Washington, D.C. DOE/EIA-0095.

———. 1992. *Financial statistics of major investor-owned electric utilities 1991.* Washington, D.C.

———. 1993. *The changing structure of the electric power industry 1970–1991.* Washington, D.C. DOE/EIA-0562.

———. Annual. *Annual outlook for U.S. electric power.* Washington, D.C. DOE/EIA-0474.

———. Annual. *Financial statistics of selected investor-owned electric utilities.* Washington, D.C. DOE/EIA-0437/1.

Federal Energy Regulatory Commission. 1988. *Opinion No. 318.* Docket EC88-2-000, Utah Power & Light and Pacific Power & Light, 45 FERC ¶61,095.

———. 1989a. *Electricity transmission: Realities, theory and policy alternatives.* Transmission Task Force Report to the Commission. Washington, D.C.

———. 1989b. *Statement of Commissioner Charles A. Trabandt on the Transmission Task Force Report to the Commission,* October 6. Washington, D.C.

———. 1990a. *Trial transcript.* Docket EC89-5-000, Southern California Edison and San Diego Gas and Electric. Washington, D.C.

———. 1990b. *Opinion 347-A.* Docket ER79-150-015, Southern California Edison, 53 FERC ¶61,101.

———. 1990c. *Initial brief of the commission trial staff.* Docket EC89-5-000, Southern California Edison and San Diego Gas and Electric. Washington, D.C.

———. 1990d. *Brief on exceptions of the commission trial staff.* Docket EC89-5-000, Southern California Edison and San Diego Gas and Electric. Washington, D.C.

———. 1990e. *Initial decision.* Docket EC89-5-000, Southern California Edison and San Diego Gas and Electric, 53 FERC ¶63,014.

———. 1990f. *Initial decision.* Docket Nos. EC90-10-000, ER90-143-000, ER90-144-000, ER90-145-000, and EL90-9-000, Northeast Utilities and Public Service of New Hampshire, 53 FERC ¶63,020.

———. 1991a. *Opinion No. 364.* Docket Nos. EC90-10-000, ER90-143-000, ER90-144-000, ER90-145-000 and EL90-9-000, Northeast Utilities and Public Service of New Hampshire, 56 FERC ¶61,269.

———. 1991b. *Order approving uncontested settlement.* Docket No. EC91-2-000, Kansas Power and Light and Kansas Gas and Electric, 56 FERC ¶61,356.

———. 1992a. *Opinion No. 364-A.* Docket Nos. EC90-10-000, ER90-10-004, ER90-143-004, ER90-144-004, ER90-145-004 and EL90-9-004, Northeast Utilities and Public Service of New Hampshire, 58 FERC ¶61,070.

———. 1992b. *Order on rate filing.* Docket No. ER91-569-000, Entergy Services.

———. 1993a. *Order on applications.* Docket Nos. EC92-21-000 and ER92-806-000, Entergy and Gulf States Utilities.

———. 1993b. *Inquiry concerning the Commission's pricing policy for transmission services provided by public utilities under the Federal Power Act: Notice of technical conference and request for comments.* Docket No. RM93-19-000, 58 Fed. Reg. 36,400, July 7.

———. 1993c. Staff discussion paper: Transmission pricing issues. Docket No. RM93-19-000, June 30. Washington, D.C.

———. 1993d. *Initial decision.* Docket Nos. EC92-21-000 and ER92-806-000, Entergy and Gulf States Utilities, 64 FERC ¶63,026.

Frankena, Mark W. 1979. *Urban transportation economics.* Toronto: Butterworths.

———. 1982. *Urban transportation financing.* Toronto: University of Toronto Press.

———. 1992. Affidavit. Appendix A to *Motion for leave to intervene...of Occidental Chemical Corp.* Federal Energy Regulatory Commission, Docket No. EC92-21-000, Entergy Services and Gulf States Utilities, September 28. Washington, D.C.

Frankena, Mark W. and Bruce M. Owen. 1992. Competitive issues in electric utility mergers. *International Merger Law* 26 (October).

———. 1993a. Antitrust analysis of electric utility mergers after the Energy Policy Act. *International Merger Law* 30 (February).

———. 1993b. Flawed reasoning. *Public Utilities Fortnightly* 131 (July 15): 25–27.

Gegax, Douglas and Kenneth Nowotny. 1993. Competition and the electric utility industry: An evaluation. *Yale Journal on Regulation* 10 (Winter): 63–87.

Gilbert, Richard J., Edward Kahn, and Matthew White. 1993. Coordination in the wholesale market: Where does it work? *Electricity Journal* 6 (October): 51-59.

Green, Douglas G. 1992. A new generation of electric utility cases emerges. *Antitrust* 7 (Fall/Winter): 28–32.

Green, Richard J. and David M. Newbery. 1992. Competition in the British electricity spot market. *Journal of Political Economy* 100 (October): 929–53.

Hahn, Robert W. and Mark V. Van Boening. 1990. An experimental examination of spot markets for electricity. *Economic Journal* 100 (December): 1073–94.

Hamilton, Neil W. and Peter R Hamilton. 1983. Duopoly in the distribution of electricity: A policy failure. *Antitrust Bulletin* 28 (Summer): 281–309.

Harris, Barry C. 1989. *Prepared direct testimony.* Federal Energy Regulatory Commission, Docket No. EC89-5-000, Southern California Edison Company and San Diego Gas and Electric, Washington, D.C.

Harris, Barry C. and Mark W. Frankena. 1992. FERC's acceptance of market-based pricing: An antitrust analysis. *Electricity Journal* 5 (June): 38–51.

Hartman, Raymond S. 1990a. The efficiency effects of electric utility mergers: Lessons from statistical cost analysis. University of California, School of Law, John M. Olin Working Papers in Law and Economics, No. 90-14.

———. 1990b. *Prepared direct testimony on revenue requirement impacts. Report on the proposed merger of Southern California Edison Company and San Diego Gas & Electric Company.* California Public Utilities Commission, Application 88-12-035, Exhibit 10,500. San Francisco.

Hartman, Raymond S., David H. Downes, and David J. Ravenscraft. 1991. A critical analysis of the proposed merger between Kansas Power and Light Company and Kansas Gas and Electric Company. Missouri Public Service Commission Staff, Case No. EM-91-213.

Hay, George A. 1993. Overview of antitrust issues in the energy industry. Paper presented to American Bar Association Annual Meeting, August 11.

Hogan, William W. 1992. Contract networks for electric power transmission. *Journal of Regulatory Economics* 4 (September): 211–42.

Holmes, A. Stewart. 1983. A review and evaluation of selected wheeling arrangements and a proposed general wheeling tariff. Federal Energy Regulatory Commission, Office of Regulatory Analysis. Mimeo.

Houston, D. 1991. Toward resolving the access issue: User-ownership of electric transmission grids. *Policy Insight* 129. Santa Monica, Calif.: Reason Foundation.

Huettner, David A. and John H. Landon. 1978. Electric utilities: Scale economies and diseconomies. *Southern Economic Journal* 44: 883–912.

Hughes, William. 1988. *Prefiled testimony.* Federal Energy Regulatory Commission, Docket EC88-2-000, Utah Power and Light Company and Pacific Power and Light, Exhibit 84. Washington, D.C.

Jacquemin, Alexis and Margaret E. Slade. 1989. Cartels, collusion, and horizontal merger. Chap. 7 in *Handbook of industrial organization*, edited by Richard Schmalensee and Robert D. Willig, vol. 1. Amsterdam: North Holland.

Jensen, Michael C. and Richard S. Ruback. 1983. The market for corporate control: The scientific evidence. *Journal of Financial Economics* 11: 5–50.

Johnson, Leland L. and David P. Reed. 1990. *Residential broadband services by telephone companies? Technology, economics, and public policy.* Santa Monica, Calif.: Rand Corp. R-3906-MF/RL.

Joskow, Paul L. 1974. Inflation and environmental concern: Structural change in the process of public utility regulation. *Journal of Law and Economics* 17 (October): 291–328.

———. 1985. Mixing regulatory and antitrust policies in the electric power industry: The price squeeze and retail market competition. Chap. 9 in *Antitrust and regulation: Essays in memory of John J. McGowan*, edited by F. M. Fisher. Cambridge, Mass.: MIT Press.

———. 1986. *The future course of competition in the electric utility industry*. Washington, D.C.: National Economic Research Associates.

———. 1989a. *Prepared direct testimony*. Federal Energy Regulatory Commission, Docket No. EC89-5-000, Southern California Edison and San Diego Gas and Electric, Exhibit 243. Washington, D.C.

———. 1989b. Regulatory failure, regulatory reform, and structural change in the electric power industry. In *Brookings papers on economic activity: Microeconomics*, edited by M. N. Baily and C. Winston, 125–99. Washington, D.C.: Brookings Institution.

———. 1990. *Prepared rebuttal testimony*. Federal Energy Regulatory Commission, Docket No. EC89-5-000, Southern California Edison and San Diego Gas and Electric, Exhibit 246. Washington, D.C.

———. 1991. The evolution of an independent power sector and competitive procurement of new generating capacity. In *Research in law and economics*, edited by R. O. Zerbe, vol. 13, 63–100. Greenwich, Conn.: JAI Press.

———. 1992. Expanding competitive opportunities in electricity generation. *Regulation* (Winter): 25–37.

———. 1993. Electricity agenda items for the new FERC. *Electricity Journal* 6 (June): 18–28.

Joskow, Paul L. and Martin L. Baughman. 1976. The future of the U.S. nuclear energy industry. *Bell Journal of Economics* 7 (Spring): 3–32.

Joskow, Paul L. and Richard Schmalensee. 1983. *Markets for power: An analysis of electric utility deregulation*. Cambridge, Mass.: MIT Press.

———. 1986. Incentive regulation for electric utilities. *Yale Journal on Regulation* 4: 1–49.

Kaserman, David L. and John W. Mayo. 1991. The measurement of vertical economies and the efficient structure of the electric utility industry. *Journal of Industrial Economics* 39 (September): 483–502.

Kimmel, Sheldon. 1987. Price correlation and market definition. Department of Justice, Antitrust Division, Discussion Paper EAG-87-8. Washington, D.C.

Kleit, Andrew N. and Richard L. Stroup. 1987. Blackout at Bonneville Power. *Regulation* (2): 30–36.

La Bella, Jeanne. 1989. The transmission access debate. *Public Power* (March–April).

Landon, John H. 1972. Electric and gas combination and economic performance. *Journal of Economics and Business* 25 (Fall): 1–13.

———. 1973. Pricing in combined gas and electric utilities: A second look. *Antitrust Bulletin* 18 (Spring): 83–98.

———. 1974. Comparative performance of combinations and separately managed utilities: Comment. *Southern Economic Journal* 41: 317–21.

Landon, John H. and John W. Wilson. 1972. An economic analysis of combination utilities. *Antitrust Bulletin* 17 (Spring): 237–68.

Loose, Verne W. and Theresa Flaim. 1980. Economies of scale and reliability: The economics of large versus small generating units. *Energy Systems and Policy* 4: 37–56.

Louis, Arthur M. 1982. The bottom line on ten big mergers. *Fortune* (May): 84–89.

Mayo, John W. 1984. Multiproduct monopoly, regulation, and firm costs. *Southern Economic Journal* 51 (July): 208–18.

Mays, Sharon. 1990. *Deposition.* California Public Utilities Commission, App. 88-12-035, Southern California Edison and San Diego Gas and Electric. San Francisco.

McCabe, K., S. Rassenti, and V. L. Smith. 1991. Experimental research on deregulation in natural gas pipeline and electric power transmission networks. In *Research in law and economics*, edited by R. O. Zerbe and V. P. Goldberg. Greenwich, Conn.: JAI Press.

Merger Standards Task Force. 1986. *Horizontal mergers: Law and policy.* Washington, D.C.: American Bar Association, Section of Antitrust Law. Monograph 12.

Morris, John R. 1992. Upstream vertical integration with automatic price adjustments. *Journal of Regulatory Economics* 4 (September): 279–87.

Morris, John R. and Gale R. Mosteller. 1991. Defining markets for merger analysis. *Antitrust Bulletin* 36 (Fall): 599–640.

Mueller, D. 1985. Mergers and market share. *Review of Economics and Statistics* 47 (May): 259–67.

Murray, T. L. 1989. *Letter to the five California Public Utilities Commissioners.* California Public Utilities Commission, App. 88-02-016, Southern California Edison reasonableness review, June 27. San Francisco. Available as Federal Energy Regulatory Commission, Docket No. EC89-5-000, Southern California Edison and San Diego Gas and Electric, Exhibit 735. Washington, D.C.

Myers, Sara S. 1989. *Reporter's transcript.* California Public Utilities Commission, App. 88-02-016, Southern California Edison reasonableness review, vol. 24, 2308-78, March 10. San Francisco.

National Independent Energy Producers. 1991. *The reliability of independent power.* Washington, D.C.

Newborn, Steven A. and Virginia L. Snider. 1992. The growing judicial acceptance of the *Merger Guidelines*. *Antitrust Law Journal* 60(3): 849–56.

Niskanen, W. A. 1992. Power to the people. *Regulation* (Winter): 11–13.

Noll, Roger G. and Bruce M. Owen. 1988. *United States v. AT&T*: An interim assessment. In *Future competition in telecommunications,* edited by J.

Hausman and S. Bradley. Cambridge, Mass.: Harvard Business School Press.

———. 1994. *United States v. AT&T*: The economic issues. In *The antitrust revolution,* rev. ed., edited by J. Kwoka and L. White. Glenview, Ill.: Scott Foresman.

Office of Technology Assessment. 1989. *Electric power wheeling and dealing: Technological considerations for increasing competition.* U.S. Congress. Washington, D.C. OTA-E-409.

Owen, Bruce M. 1970. Monopoly pricing in combined gas and electric utilities. *Antitrust Bulletin* 15 (Winter): 713–26.

———. 1973. Pricing in combined gas and electric utilities: Reply. *Antitrust Bulletin* 18 (Spring): 99.

———. 1989. *Prepared direct testimony*. Federal Energy Regulatory Commission, Docket No. EC89-5-000, Southern California Edison and San Diego Gas and Electric, Exhibit 710. Washington, D.C.

———. 1990a. *Surrebuttal testimony and exhibits to applicant witness Jurewitz of Bruce M. Owen.* Federal Energy Regulatory Commission, Docket No. EC89-5-000, Southern California Edison and San Diego Gas and Electric, Exhibit 1067. Washington, D.C.

———. 1990b. *Surrebuttal testimony and exhibits to applicant witness Pace.* Federal Energy Regulatory Commission, Docket EC89-5-000, Southern California Edison and San Diego Gas and Electric, Exhibit 1070. Washington, D.C.

———. 1993a. *Affidavit. Chesapeake and Potomac Tel. Co. of Va. et al. v. United States*, Civil Action No. 92-1751-A (E.D. Va), May 20.

———. 1993b. *Affidavit. Chesapeake and Potomac Tel. Co. of Va. et al. v. United States*, Civil Action No. 92-1751-A (E.D. Va), June 9.

Owen, Bruce M. and Peter R. Greenhalgh. 1986. Competitive considerations in cable television franchising. *Contemporary Policy Issues* 4 (April): 69–79.

Owen, Bruce M. and Steven S. Wildman. 1992. *Video economics*. Cambridge, Mass.: Harvard University Press.

Pace, Joe D. 1972. The relative performance of combination gas-electric utilities. *Antitrust Bulletin* 17 (Summer): 519–65.

———. 1989. *Prepared direct testimony*. Federal Energy Regulatory Commission, Docket No. EC89-5-000, Southern California Edison and San Diego Gas and Electric, Exhibit 233. Washington, D.C.

———. 1990. *Prepared rebuttal testimony*, Federal Energy Regulatory Commission, Docket No. EC89-5-000, Southern California Edison and San Diego Gas and Electric, Exhibit 236. Washington, D.C.

Pace, Joe D. and John H. Landon. 1982. Introducing competition into the electric utility industry: An economic appraisal. *Energy Law Journal* 3(2): 1–65.

Pautler, Paul A. 1983. A review of the economic basis for broad-based horizontal-merger policy. *Antitrust Bulletin* 28 (Fall): 571–651.

Plummer, James and Susan Troppmann, eds. 1990. *Competition in electricity: New markets and new structures*. Arlington, Va.: Public Utilities Reports.

Pratt, Robert W. 1992. The "sophisticated buyer" defense in merger litigation gains momentum. *Antitrust* 6 (Spring): 9–13.

Primeaux, Walter J., Jr. 1974. A duopoly in electricity: Competition in a "natural monopoly." *Quarterly Review of Economics and Business* 14 (Summer): 65–73.

———. 1975a. A reexamination of the monopoly market structure for electric utilities. Chap. 6 in *Promoting competition in regulated markets*, edited by A. Phillips. Washington, D. C.: Brookings Institution.

———. 1975b. The decline in electric utility competition. *Land Economics* 51 (May): 144–48.

Ravenscraft, D. and F.M. Scherer. 1989. The profitability of mergers. *International Journal of Industrial Organization* 7: 101–16.

Reynolds, Stanley S. 1990. Cost sharing and competition among daily newspapers. University of Arizona, Department of Economics.

Rill, James F. 1991. *Statement concerning S. 173,* U.S. Senate, Comm. on Commerce, Science, and Transp., Subcomm. on Communications, Feb. 28.

Roberts, Mark J. 1986. Economies of density and size in the production and delivery of electric power. *Land Economics* 4 (November): 378–87.

Robinson, Jim. 1992. The KPL/KGE merger: A regulator's view. *Public Utilities Fortnightly* 129 (May 15): 18–19.

Ruff, L. E. 1988. Least-cost planning and demand-side management. *Public Utilities Fortnightly* 121 (April 28): 19–26.

Rule, Charles F. 1988. Antitrust and bottleneck monopolies: The lessons of the AT&T decree. Remarks before the Brookings Institution Seminar on Developments in Telecommunications Policy, Washington, D.C., Oct. 5.

Saacks, Jerry J. 1992. *Prepared direct testimony and exhibits*. Federal Energy Regulatory Commission, Docket Nos. EC92-21-000 and ER92-806-000, Entergy and Gulf States Utilities, Exhibit Nos. JJS-1 to JJS-7. Washington, D.C.

Scheffman, David T. 1993. Ten years of *Merger Guidelines*: A retrospective, critique, and prediction. *Review of Industrial Organization* 8(2): 173–89.

Scheffman, David T. and Pablo T. Spiller. 1987. Geographic market definition under the U.S. Department of Justice *Merger Guidelines*. *Journal of Law and Economics* 30 (April): 123–47.

Scherer, F. M. and David Ross. 1990. *Industrial market structure and economic performance*. 3d ed. Boston: Houghton Mifflin.

Schmalensee, Richard and Bennett W. Golub. 1984. Estimating effective concentration in deregulated wholesale electricity markets. *Rand Journal of Economics* 15 (Spring): 12–26.

Schroeder, Christopher H., Lyna L. Wiggins and Daniel T. Wormhoudt. 1981. Flexibility of scale in large conventional coal-fired plants. *Energy Policy* (June): 127–35.

Shleifer, Andrei. 1985. A theory of yardstick competition. *Rand Journal of Economics* 16 (Autumn): 319–27.

Simons, Nelson W., Len L. Garver, and Rana Mukerji. 1993. Transmission constrained production costing: A key to pricing transmission access. *Public Utilities Fortnightly* 131 (February 1): 52–55.

Sing, Merrile. 1987. Are combination gas and electric utilities multiproduct natural monopolies? *Review of Economics and Statistics* 69 (August): 392–98.

Smith, Vernon L. 1987. Currents of competition in electricity markets. *Regulation* (2): 23–29.

———. 1988. Electric power deregulation: Background and prospects. *Contemporary Policy Issues*: 14–24.

———. 1991. Can electric power—a "natural monopoly"—be deregulated? University of Arizona, Economic Science Laboratory. Mimeo.

Southern California Edison. 1989. *QF reasonableness general policy testimony*. California Public Utilities Commission. San Francisco. Available as Federal Energy Regulatory Commission, Docket No. EC89-5-000, Southern California Edison and San Diego Gas and Electric, Exhibit 741. Washington, D.C.

Southern California Edison and San Diego Gas and Electric. 1990. *Brief on exceptions of applicants*. Federal Energy Regulatory Commission, Docket EC89-5-000. Washington, D. C.

Stalon, Charles G. 1993. The FERC: Designated driver and head herdsman for electric industry restructuring. *Electricity Journal* 6 (June): 29–35.

Stevenson, Rodney. 1982. X-inefficiency and interfirm rivalry: Evidence from the electric utility industry. *Land Economics* 58 (February): 52–66.

Stockum, Steve. 1993. The efficiencies defense for horizontal mergers: What is the government's standard? *Antitrust Law Journal* 61 (Spring): 829–55.

Stoner, Robert D. 1993. *Prepared direct testimony*. Federal Energy Regulatory Commission, Docket Nos. EC92-21-000 and ER92-806-000, Entergy and Gulf States Utilities, Exhibit Nos. OCC-1 through OCC-13. Washington, D.C.

Strategic Decisions Group. 1991. *Western Systems Power Pool assessment: Final report*. Menlo Park, Calif.

Swidler, Joseph C. 1991. An unthinkably horrible situation. *Public Utilities Fortnightly* 128 (September 15): 14–18, 40.

Taylor, Gordon T. C. 1989. *Direct testimony*. Federal Energy Regulatory Commission, Docket No. EC89-5-000, Southern California Edison and San Diego Gas and Electric, Exhibit 628. Washington, D.C.

Tenenbaum, Bernard, and Stephen Henderson. 1991. Market-based pricing of wholesale electric services. *Electricity Journal* 4 (December): 30–45.

Tirole, Jean. 1988. *The theory of industrial organization*. Cambridge, Mass.: MIT Press.

U.S. Department of Commerce, Bureau of the Census. 1988. *1986 Annual survey of manufactures: Statistics for industry groups and industries*. Washington, D.C. M86(AS)-1.

U.S. Department of Energy. 1991/1992. *Analysis of options to amend the Public Utility Holding Company Act of 1935*. National Energy Strategy, Technical Annex 1. Washington, D.C. DOE/S-0084P.

U.S. Department of Justice. 1968. *Merger Guidelines*. Washington, D.C. Reprinted in 4 Trade Reg. Rep. (CCH) ¶13,101.

———. 1976. *Report of the Attorney General on the applications of LOOP, Inc. and Seadock, Inc. for deepwater port licenses*. Antitrust Division. Washington, D.C.

———. 1982. *Competition in the coal industry*. Antitrust Division. Washington, D.C.

———. 1984a. *Competition in the oil pipeline industry: A preliminary report*. Antitrust Division. Washington, D.C.

———. 1984b. *Merger Guidelines*. Washington, D.C. Reprinted in 4 Trade Reg. Rep. (CCH) ¶13,103.

———. 1990. *Brief of the U.S. Department of Justice on exceptions to the initial decision of the administrative law judge*. Federal Energy Regulatory Commission, Docket No. EC89-5-000, Southern California Edison and San Diego Gas and Electric. Washington, D. C.

U.S. Department of Justice and Federal Trade Commission. 1992. *Horizontal Merger Guidelines*. Washington, D.C. Repinted in 4 Trade Reg. Rep. (CCH) ¶13,104.

Uri, Noel D. and Malcolm B. Coate. 1987. The Department of Justice Merger Guidelines: The search for empirical support. *International Review of Law and Economics* 7: 113–20.

Werden, Gregory J. 1981. The use and misuse of shipments data in defining geographic markets. *Antitrust Bulletin* 26 (Winter): 71937.

———. 1983. Market delineation and the Justice Department's Merger Guidelines. *Duke Law Journal* (June): 514–79.

Whalley, J. 1989. After the Herfindahls are counted: Assessment of entry and efficiencies in merger enforcement by the Department of Justice. U.S. Department of Justice, Antitrust Division. Washington, D.C.

Wildman, Steven S. and Margaret E. Guerin-Calvert. 1991. Electronic services networks: Functions, structures, and public policy. In *Electronic services networks*, edited by M. E. Guerin-Calvert and S. S. Wildman, 3–21. New York: Praeger.

Williamson, Oliver E. 1975. *Markets and hierarchies: Analysis and antitrust implications*. New York: The Free Press/MacMillan.

———. 1976. Franchise bidding for natural monopolies—in general and with respect to CATV. *Bell Journal of Economics* 7 (Spring): 73–104.

Subject Index

Author Index

About the Authors

MARK W. FRANKENA is Senior Economist with Economists Incorporated, where he has been employed since 1988. Between 1982 and 1988, Dr. Frankena was employed by the U.S. Federal Trade Commission, Bureau of Economics, where he supervised antitrust analysis.

BRUCE M. OWEN has been President of Economists Incorporated since 1981 and Visiting Professor of Economics at Stanford University's Washington, D.C., campus since 1989. He was Chief Economist of the Antitrust Division of the U.S. Department of Justice from 1979 through 1981; prior to this, Dr. Owen was a faculty member at Duke and Stanford universities.